SpringerBriefs in Applied Sciences and Technology

Series editor

Janusz Kacprzyk, Polish Academy of Sciences, Systems Research Institute, Warsaw, Poland

SpringerBriefs present concise summaries of cutting-edge research and practical applications across a wide spectrum of fields. Featuring compact volumes of 50–125 pages, the series covers a range of content from professional to academic.

Typical publications can be:

- A timely report of state-of-the art methods
- An introduction to or a manual for the application of mathematical or computer techniques
- A bridge between new research results, as published in journal articles
- A snapshot of a hot or emerging topic
- An in-depth case study
- A presentation of core concepts that students must understand in order to make independent contributions

SpringerBriefs are characterized by fast, global electronic dissemination, standard publishing contracts, standardized manuscript preparation and formatting guidelines, and expedited production schedules.

On the one hand, **SpringerBriefs in Applied Sciences and Technology** are devoted to the publication of fundamentals and applications within the different classical engineering disciplines as well as in interdisciplinary fields that recently emerged between these areas. On the other hand, as the boundary separating fundamental research and applied technology is more and more dissolving, this series is particularly open to trans-disciplinary topics between fundamental science and engineering.

Indexed by EI-Compendex and Springerlink.

More information about this series at http://www.springer.com/series/8884

Zahra Trad · Abdelwahed Barkaoui
Moez Chafra · João Manuel R. S. Tavares

FEM Analysis of the Human Knee Joint

A Review

 Springer

Zahra Trad
LR-11-ES19 Laboratoire de Mécanique
  Appliquée et Ingénierie (LR-MAI), Ecole
  Nationale d'Ingénieurs de Tunis
Université de Tunis El Manar
Tunis
Tunisie

Abdelwahed Barkaoui
LR-11-ES19 Laboratoire de Mécanique
  Appliquée et Ingénierie (LR-MAI), Ecole
  Nationale d'Ingénieurs de Tunis
Université de Tunis El Manar
Tunis
Tunisie

Moez Chafra
Laboratoire de Systèmes et de Mécanique
  Appliquée
Ecole Polytechnique de Tunis
Tunis
Tunisie

and

IPEIEM
Université de Tunis El Manar
Tunis
Tunisie

João Manuel R. S. Tavares
Instituto de Ciência e Inovação em
  Engenharia Mecânica e Engenharia
  Industrial, Departamento de Engenharia
  Mecânica, Faculdade de Engenharia
Universidade do Porto
Porto
Portugal

ISSN 2191-530X   ISSN 2191-5318 (electronic)
SpringerBriefs in Applied Sciences and Technology
ISBN 978-3-319-74157-4   ISBN 978-3-319-74158-1   (eBook)
https://doi.org/10.1007/978-3-319-74158-1

Library of Congress Control Number: 2017963858

© The Author(s), under exclusive licence to Springer International Publishing AG, part of Springer Nature 2018, corrected publication June 2018
This work is subject to copyright. All rights are reserved by the Publisher, whether the whole or part of the material is concerned, specifically the rights of translation, reprinting, reuse of illustrations, recitation, broadcasting, reproduction on microfilms or in any other physical way, and transmission or information storage and retrieval, electronic adaptation, computer software, or by similar or dissimilar methodology now known or hereafter developed.
The use of general descriptive names, registered names, trademarks, service marks, etc. in this publication does not imply, even in the absence of a specific statement, that such names are exempt from the relevant protective laws and regulations and therefore free for general use.
The publisher, the authors and the editors are safe to assume that the advice and information in this book are believed to be true and accurate at the date of publication. Neither the publisher nor the authors or the editors give a warranty, express or implied, with respect to the material contained herein or for any errors or omissions that may have been made. The publisher remains neutral with regard to jurisdictional claims in published maps and institutional affiliations.

Printed on acid-free paper

This Springer imprint is published by the registered company Springer International Publishing AG part of Springer Nature
The registered company address is: Gewerbestrasse 11, 6330 Cham, Switzerland

Without a curious mind, what would we be?
Such is the beauty and nobility
of science: an endless desire to push back
the frontiers of knowledge, to hunt down
the secrets of matter and life without
any preconceived idea of the possible
consequences.

Marie Curie

Everything I am or aspire to become, I owe it to my mother, my angel.

These pages are a way to say thank you.

I will always thank you.

I will always love you.

With love,

Zahra Trad

Acknowledgements

The authors gratefully acknowledge the funding of Project NORTE-01-0145-FEDER-000022—SciTech—Science and Technology for Competitive and Sustainable Industries, co-financed by the "Programa Operacional Regional do Norte" (NORTE2020), through the "Fundo Europeu de Desenvolvimento Regional" (FEDER).

Contents

Introduction

Biomechanics is the study of the mechanical laws relating to the movement and/or structure of living organisms. It often refers to the study of human body interaction with the environment under different conditions. Particularly, the knee joint is the largest and most complex and stressed joint in the human body, comprised of both the femoro-patellar and tibiofemoral joints. The important structure of the knee joint is divided into the following main tissues. The soft tissues include ligaments, tendons, menisci, muscles, and articular cartilages. The bony structure (hard tissues) includes the femur, tibia, fibula, and patella. Owing to the bearing of very high loads, the knee structure permits active movement for daily activities such as walking, running, sitting, and kneeling, which produce complex mechanical responses under the loads that occur in everyday life (Moglo and Shirazi-Adl 2003; Sahu and Kaviti 2016; Gatchel et al. 2007; Kiapour et al. 2014; Biščević et al. 2005; Zaffagnini et al. 2013).

However, excessive joint loading on the articular cartilage, as well as age, weight bearing, meniscectomy, alignment, and joint injury, may cause various types of damage and lead to the initiation and progression of knee osteoarthritis (OA). Clinical studies have revealed that an increased risk in knee OA progression with abnormal tibiofemoral alignment most likely results from the increase of mechanical loading at the knee joint (Andriacchi et al. 2004; Sharma et al. 2001; Felson and Zhang 1998; Felson et al. 2000; Sowers 2001; Cooper et al. 2000; Sharma 2001; Cerejo et al. 2002; Englund and Lohmander 2004; Felson et al. 2004; Griffin and Guilak 2005; Kraus et al. 2013; Kubo et al. 2009; Arokoski et al. 2000).

The role of the loading axis in the progression of knee OA has been well recognized. Clinically, the interventions that modify the excessive forces related to the position of the loading axis acting on damaged tissues, such as articular cartilage, have been applied in relieving OA symptoms (Akizuki et al. 2008). For knees afflicted by medial OA progression, high tibial osteotomy (HTO) can be an effective adjunct to conservative management of interruption of the disease's progression (Schallberger et al. 2011; Esenkaya and Elmali 2006). By correcting the hip–knee–ankle (HKA) angle, HTO shifts the mechanical load from the arthritic medial compartment to the lateral compartment with relatively healthy cartilage.

Thereby, a specific analysis of the stress and strain distribution, contact kinematics, and loading at the knee should be predicted to quantify soft tissue mechanics under different loading conditions. Finite element analysis (FEA) is a promising method and a valuable and impressive tool used to model biomechanical structures in favor of overcoming the challenge of performing direct measurement of forces, stresses, and strains in the bones and soft tissues at different joints by developing three-dimensional (3D) finite element (FE) models of the knee joint. Computational modeling, such as FEA, is the final part of the framework that integrates subject-specific joint geometry, kinematics, and loading to obtain subject-specific data on the stresses and strains within the knee joint (Andriacchi et al. 2004; Fernandez and Pandy 2006; Besier et al. 2005). This biomechanical information can be extremely useful in a clinical context (Kiapour et al. 2014; Herrera et al. 2012; Kluess et al. 2010; Peña et al. 2006; Gardiner and Weiss 2003).

The present manuscript summarizes the current biomechanical and clinical knowledge on the human knee joint. It provides a review of FEA of this joint and brings to light the significance of this method for this type of analysis.

In the first section, the different techniques used to model the knee joint, such as the material properties assigned to each structure of the joint (bones, cartilage, menisci and ligament), are discussed.

The second section highlights the main role of FEA in predicting the mechanical behavior of articular cartilage, ligaments, and menisci. It is carried out to enumerate biomechanical studies that have investigated the influence of material properties, loads and boundary conditions, ligament injury, meniscectomy, cartilage disease, and other parameters on the stress and strain distribution in the knee joint. A summary of the FEA studies on the HTO procedure is presented at the end of this section.

Owing to the lack of consensus on the optimal correction angle for maximizing osteotomy survival and postoperative knee function, the third section quantifies both clinical and biomechanical studies on optimizing the correction angle.

The last section presents the main conclusions and recommendations of critical areas for further studies.

References

Akizuki, S., et al. (2008). The long-term outcome of high tibial osteotomy. *Bone & Joint Journal, 90*(5), 592–596.

Andriacchi, T. P., et al. (2004). A framework for the in vivo pathomechanics of osteoarthritis at the knee. *Annals of Biomedical Engineering, 32*(3), 447–457.

Arokoski, J. P. A., et al. (2000). Normal and pathological adaptations of articular cartilage to joint loading. *Scandinavian Journal of Medicine & Science in Sports, 10*(4), 186–198.

Besier, T. F., et al. (2005). A modeling framework to estimate patellofemoral joint cartilage stress in vivo. *Medicine and Science in Sports and Exercise, 37*(11), 1924–1930.

Biščević, M., Hebibović, M., & Smrke, D. (2005). Variations of femoral condyle shape. *Collegium Antropologicum, 29*(2), 409–414.

Cerejo, R., et al. (2002). The influence of alignment on risk of knee osteoarthritis progression according to baseline stage of disease. *Arthritis & Rheumatism, 46*(10), 2632–2636.

Cooper, C., et al. (2000). Risk factors for the incidence and progression of radiographic knee osteoarthritis. *Arthritis & Rheumatism, 43*(5), 995–1000.

Englund, M., & Lohmander, L. S. (2004). Risk factors for symptomatic knee osteoarthritis fifteen to twenty-two years after meniscectomy. *Arthritis & Rheumatism, 50*(9), 2811–2819.

Esenkaya, I., & Elmali, N. (2006). Proximal tibia medial open-wedge osteotomy using plates with wedges: Early results in 58 cases. *Knee Surgery, Sports Traumatology, Arthroscopy, 14*(10), 955–961.

Felson, D. T., & Zhang, Y. (1998). An update on the epidemiology of knee and hip osteoarthritis with a view to prevention. *Arthritis & Rheumatology, 41*(8), 1343–1355.

Felson, D. T., et al. (2000). Osteoarthritis: New insights. Part 1: The disease and its risk factors. *Annals of Internal Medicine, 133*(8), 635–646.

Felson, D. T., et al. (2004). The effect of body weight on progression of knee osteoarthritis is dependent on alignment. *Arthritis & Rheumatism, 50*(12), 3904–3909.

Fernandez, J. W., & Pandy, M. G. (2006). Integrating modelling and experiments to assess dynamic musculoskeletal function in humans. *Experimental Physiology, 91*(2), 371–382.

Gardiner, J. C., & Weiss, J. A. (2003). Subject-specific finite element analysis of the human medial collateral ligament during valgus knee loading. *Journal of Orthopaedic Research, 21*(6), 1098–1106.

Gatchel, R. J., et al. (2007). The biopsychosocial approach to chronic pain: Scientific advances and future directions. *Psychological Bulletin, 133*(4), 581.

Griffin, T. M., & Guilak, F. (2005). The role of mechanical loading in the onset and progression of osteoarthritis. *Exercise and Sport Sciences Reviews, 33*(4), 195–200.

Herrera, A., et al. (2012). Applications of finite element simulation in orthopedic and trauma surgery. *World Journal of Orthopedics, 3*(4), 25.

Kiapour, A., et al. (2014). Finite element model of the knee for investigation of injury mechanisms: Development and validation. *Journal of Biomechanical Engineering, 136*(1), 11002.

Kiapour, A. M., et al. (2014). The effect of ligament modeling technique on knee joint kinematics: A finite element study. *Applied Mathematics, 4*(5A), 91.

Kluess, D., et al. (2010). *Finite element analysis in orthopaedic biomechanics*. INTECH Open Access Publisher.

Kraus, V. B., et al. (2013). High prevalence of contralateral ankle abnormalities in association with knee osteoarthritis and malalignment. *Osteoarthritis and Cartilage, 21*(11), 1693–1699.

Kubo, M., et al. (2009). Chondroitin sulfate for the treatment of hip and knee osteoarthritis: Current status and future trends. *Life Sciences, 85*(13), 477–483.

Moglo, K. E., & Shirazi-Adl, A. (2003). Biomechanics of passive knee joint in drawer: Load transmission in intact and ACL-deficient joints. *The Knee, 10*(3), 265–276.

Peña, E., et al. (2006). A three-dimensional finite element analysis of the combined behavior of ligaments and menisci in the healthy human knee joint. *Journal of Biomechanics, 39*(9), 1686–1701.

Sahu, N. K., & Kaviti, A. K. (2016). A Review of use FEM Techniques in modeling of human knee joint. *Journal of Biomimetics, Biomaterials and Biomedical Engineering*, 14–25 (Trans Tech Publications).

Schallberger, A., et al. (2011). High tibial valgus osteotomy in unicompartmental medial osteoarthritis of the knee: a retrospective follow-up study over 13–21 years. *Knee Surgery, Sports Traumatology, Arthroscopy, 19*(1), 122–127.

Sharma, L., et al. (2001). The role of knee alignment in disease progression and functional decline in knee osteoarthritis. *JAMA, 286*(2), 188–195.

Sharma, L. (2001). Local factors in osteoarthritis. *Current Opinion in Rheumatology, 13*(5), 441–446.

Sowers, M. (2001). Epidemiology of risk factors for osteoarthritis: Systemic factors. *Current Opinion in Rheumatology, 13*(5), 447–451.

Zaffagnini, S., et al. (2013). A standardized technique in performing pivot-shift test on the knee joint provided more consistent acceleration curve shape, allowing to highlight side-to-side differences. *Arthroscopy, 29*(10), e175.

Abstract

In recent years, numerous scientific investigations have studied the anatomical, biomechanical, and functional role of the structures involved in the human knee joint. The finite element method (FEM) has been seen as an interesting tool for studying and simulating biosystems; in particular, FEM has been extensively used to analyze the knee joint and various types of knee disease and rehabilitation procedure, such as high tibial osteotomy (HTO). The aim of this manuscript is to present a review of FEM Analysis of the Human Knee Joint and HTO knee surgery and discuss the adequacy of this computational tool for this type of biomedical application. Hence, various studies addressing the knee joint based on finite element analysis (FEA) are reviewed, and an overview of clinical and biomechanical studies addressing the optimization of the correction angle of postoperative knee surgery is provided.

Chapter 1
Finite Element Models of the Knee Joint

Geometry, material properties and loading and boundary conditions are important aspects in FEA modeling of the knee. Many assumptions and concessions must be made while considering these aspects, as well as the computational time. For example, it is very computationally expensive to model the cartilage or the menisci as poroelastic materials. Moreover, modeling the cartilage as three layers (superficial, middle and deep) is very difficult using 3D models, but can be easily accomplished using 2D models, and simulations can therefore be completely performed.

1.1 Knee Joint Model Geometries

Understanding the biomechanical behavior of human diarthrodial joints, including the knee joint, is essential for developing better diagnostic techniques, fostering objective comparisons of alternate surgical interventions, and facilitating the functional engineering of tissue replacements. Therefore, several mathematical and finite element models of the knee joint are essential to obtain a more complete understanding of the biomechanical behavior and an evaluation of the surgical and diagnostic procedures. Recent developments show that 3D mathematical models of the knee joint develop into powerful tools for the design of artificial joints and functional analysis of the knee (Essinger et al. 1989).

For these reasons, mathematical models of the human knee joint have been used to predict the biomechanical behavior of individual tissue structures (Abdel-Rahman and Hefzy 1993; Andriacchi et al. 1983; Bach et al. 1992; Bendjaballah et al. 1995; Beynnon et al. 1996; Blankevoort and Huiskes 1996a, b; Crowninshield et al. 1976; Gibson et al. 1986; O'Connor 1993; Shelburne and Pandy 1997; Turner and Engin 1993; Wismans et al. 1980; Zavatsky and O'Connor 1993).

For example, Wismans et al. (1980) developed a 3D mathematical model to study the relative motions and forces in the human knee joint. The tibia and the

© The Author(s), under exclusive licence to Springer International Publishing AG, part of Springer Nature 2018

Z. Trad et al., *FEM Analysis of the Human Knee Joint*, SpringerBriefs in Applied Sciences and Technology, https://doi.org/10.1007/978-3-319-74158-1_1

femur were considered as rigid bodies that are in contact with the ligament at two points. The ligaments were modeled as nonlinear elastic springs. The patella and the menisci were not included in the model. The articular surfaces were represented by polynomials in space with neglected friction. The constant external loads acting on the femur represent the forces induced by muscles and weight. As a result, if the joint moves from extension to flexion, a tendency for the tibia to perform an internal rotation was observed. However, it was also found that the magnitude of this rotation strongly depends on the magnitude of the external load acting on the joint. The general shape of the curves characterizing the relation between a torque about the tibial longitudinal axis and the resulting internal or external rotation at different flexion-extension angles has been found to be quite similar to that of experimental curves (Markolf et al. 1976). Finally, the internal-external free range of motion is found to be relatively small in extension and to increase strongly in flexion. Therefore, this model is considered as a kinematical model, as well as a model for the joint forces, which allows for the determination of the relative position of the tibia and the femur as a function of an external load and the flexion-extension angle. The following valuable perspectives related to several problems of a more practical nature could be discussed:

– The contact stresses in the joint can be studied by integrating the contact forces with the geometrical and material properties of the articular surfaces and the menisci. This is essential for insight into the pathogenesis of OA.
– Some aspects of the replacement of a ligament by an artificial component could be evaluated.
– Comparison could be made between the different types of prosthesis in order to evaluate certain aspects of correcting positioning and alignment of these prostheses.

Thus, Andriacchi et al. (1983) developed a 3D mathematical model of the ligamentous knee joint, including the proximal tibia, distal femur, soft tissue structure and the contacting surfaces of the medial and lateral condyles. The menisci and the capsule were not modeled. The main goal was to study the mechanical response of the knee joint. The bones of the model were considered as rigid bodies, while the soft tissue structures were represented by spring- and beam-type elements. The direct stiffness approach from structural mechanics, as well as an incremental linearization procedure for the geometric and material non-linearities, was used in the model. Results of this work revealed that the load-displacement response of the knee is highly dependent on constraints to coupled degrees of freedom. This finding may have an important consideration when interpreting the results of standard laxity tests at the knee, which, by their nature, may impose constraints on motion.

Further study reported by Essinger et al. (1989) presented a 3D model used as input for a computer program in order to evaluate the mechanical behavior during flexion of a condylar-type knee prosthesis (C-TKA). Similar to the model of Wismans et al. (1980), the menisci and the capsule were not modeled, while a simplified patellofemoral joint was included. The kinematics of the joint, the motion of the center of contact, the quadriceps forces, the pressure distribution on the tibial plateau, and the ligament lengths and forces between 0 and 120° of flexion were generated based on the total energy minimization principle. The program simulates the kinematics of the joint in a similar manner to that used in the experimental jig of Reuben et al. (1986), Kurosawa et al. (1985) and Rovick et al. (1986). Their results revealed that the motion of the C-TKA is strongly related to the geometry of the prosthetic articular surfaces. The motion pattern was indeed one criterion for the design of this type of prosthesis, which is strongly dependent upon the type of patient. According to this work, the proposed program can be a useful tool for the design of new components, as it may be used as a preliminary evaluation process. However, it cannot serve as a substitute for in vitro and clinical studies because of the lack of joint laxity evaluations.

Later, Blankevoort et al. (1991) studied the effect of articular contact on the passive motion characteristics in order to experimentally obtain joint kinematics through a 3D mathematical model of the knee joint. In their study, two different mathematical contact descriptions (rigid contact vs. deformable contact) were compared for this purpose. The description of deformable contact is based on a simplified theory for contact of thin elastic material or as a non-linear elastic material. The contact descriptions were introduced in a mathematical model of the knee. The geometry of the articular surfaces and the locations of the ligament insertions were obtained from a joint specimen on which experimentally determined kinematic data were available and were used as input for the model. The ligaments were considered as non-linear elastic line elements. The mechanical properties of the ligaments and the articular cartilage were derived from date from the literature (Blankevoort et al. 1988; Kempson 1980; Mow et al. 1982; Walker and Hajek 1972; Roth 1977). Principal results of this work showed that for simulation of the passive motion characteristics of the knee, the simplified description for contact of a thin linear elastic layer on a rigid foundation is a valid approach when aiming at the study of the motion characteristics for moderate loading conditions. Hence, with deformable contact in the knee model, geometric conformity between the surfaces can be modelled as opposed to rigid contact, which assumed only point contact.

While mathematical models of the knee joint can be useful in predicting the forces and stresses in individual knee structures and other parameters such as knee kinematics, validation of such a model always presents a challenge. In other words, it is difficult to construct computational knee models that predict joint motion and forces comparable to those obtained through experiments. A literature survey revealed that both 2D and 3D models have been used in FEA simulations to investigate contact stresses and strains at the knee joint.

Vaziri et al. (2008) presented, in their study, a typical 2D axi-symmetric knee model that has been used in numerous 2D studies (Herzog 2004; Un 2001; Wilson et al. 2005; Dar and Aspden 2003; Wilson et al. 2003, 2004; Vadher et al. 2006; Donzelli et al. 1999). The geometry used in 2D represents axi-symmetric approximations of 3D contact geometries found in the knee and still provide valuable insight into knee mechanics and the initiation and progression of OA.

Because of the difficulty and time-consuming nature of creating the geometries of the 3D knee models, the most commonly used process involves manually digitizing an array of 2D images to create 3D surfaces of the bones, cartilage and menisci. In addition, ligaments are modeled, as they add important static and dynamic stability to the knee joint. Generally, in FEA models, the tendons and muscles are not included, loads and moments are applied at specific points and contact contribution by the tendons, and the muscles are neglected.

The earliest 3D model geometry was based on geometric measurements of the knee. Wismans et al. (1980) developed a 3D analytical model of the knee joint using geometric data points of the articular surfaces of the tibia and femur to obtain the parameters for surface polynomials using the least squares method. The ligaments and capsule are represented by a number of non-linear springs, with material properties selected from the literature (Trent et al. 1976; Wang et al. 1973). As result, the predictions of the model agree well with experiments described in the literature (Bartel et al. 1977; Zuppinger 1904; Markolf et al. 1976; Hsieh and Walker 1976; Wang and Walker 1974; Brantigan and Voshell 1941).

Pandy et al. (Pandy and Sasaki 1998; Pandy et al. 1997) developed a 3D knee model in order to study ligament function during anterior-posterior draw, axial rotation, and isometric contractions of the extensor and flexor muscles. The shapes of the articulating surfaces of the distal femur, proximal tibia, and patella were created using fitting polynomials to digitize data reported for 23 cadaver knees (Garg and Walker 1990).

Another method that has been used involves acquiring the different knee structures through Magnetic Resonance Imaging (MRI) or Computed Tomography

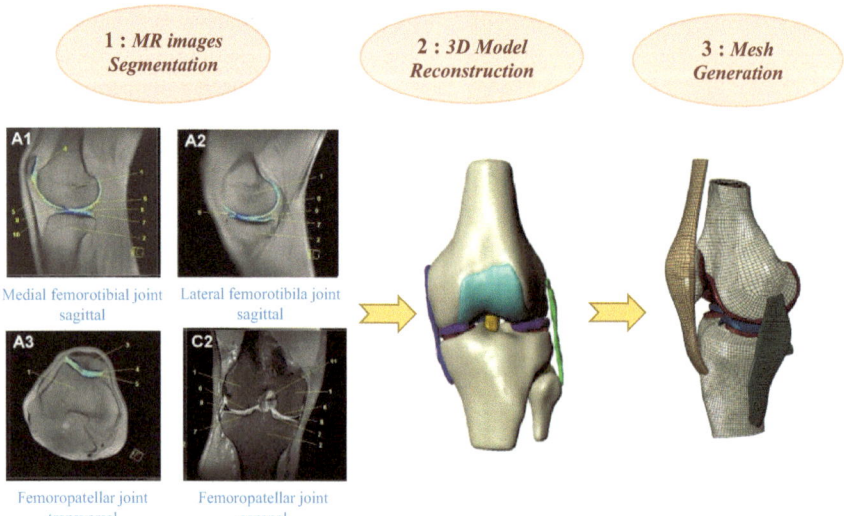

Fig. 1.1 Knee joint finite element model development steps [image 1 taken from (Schütz et al. 2012); image 2 taken from (Zheng 2014); and image 3 taken from (Wang et al. 2014)]

(CT) to obtain the 3D geometry (Fig. 1.1). Haut et al. (1997) and Donahue et al. (2002, 2003) manually acquired CT images of the bones in the knee of a cadaver specimen. The soft tissue was modeled using a laser-based 3D coordinate digitizing system. The images were used to develop 3D geometries for FEA simulations. This solution produced highly accurate models. Unfortunately, the laser-based system does not have the ability to measure live human knee joint subjects, because it can only digitize the knee joint after the skin and muscle architecture have been removed from the bones.

Since the '90s, most published studies have used MRI to digitize both the hard bone tissue as well as the soft tissue at the knee. Generally, these studies used CT and MRI scans of cadaveric human knee specimens (Ali et al. 2016; Donahue et al. 2002; Gardiner and Weiss 2003; Guess et al. 2010; Li et al. 1999a; Mina et al. 2008; Mootanah et al. 2014; Yao et al. 2006b; Song et al. 2004) or a specific volunteer (Mononen et al. 2012; Peña et al. 2005a, 2006a, b, 2007, 2008; Penrose et al. 2002; Blecha et al. 2005; Halonen et al. 2013; Zheng et al. 2014; Zheng 2014; Yang et al. 2010a, b; Wang et al. 2014; Kiapour et al. 2014; Miyoshi et al. 2002a).

With the newly developed robotic/UFS system (Rudy et al. 1996) as a testing platform, Li et al. (1999a) developed the first 3D, FE tibio-femoral knee joint model in order to predict knee kinematics and forces in ligaments in response to external loads. This model was experimentally validated by comparing it with previous experimental studies (Kanamori et al. 1998; Markolf et al. 1981, 1993, 1995; Wascher et al. 1993; Woo et al. 1998; Blankevoort et al. 1991; Blankevoort and Huiskes 1991, 1996a, b; Bendjaballah et al. 1995; Andriacchi et al. 1983). At that time, the geometry of the knee joint was obtained from an MRI of a cadaveric knee specimen (Fig. 1.2), in which both soft and hard tissues were included. Cartilage was modeled as an elastic material, menisci were simulated by equivalent-resistance springs, and ligaments were represented as nonlinear elastic springs. The mechanical properties of the ligaments and the articular cartilage were derived from data taken from the literature (Blankevoort and Huiskes 1996a, b; Blankevoort et al. 1991; Butler et al. 1986). The same specimen was tested biomechanically using a

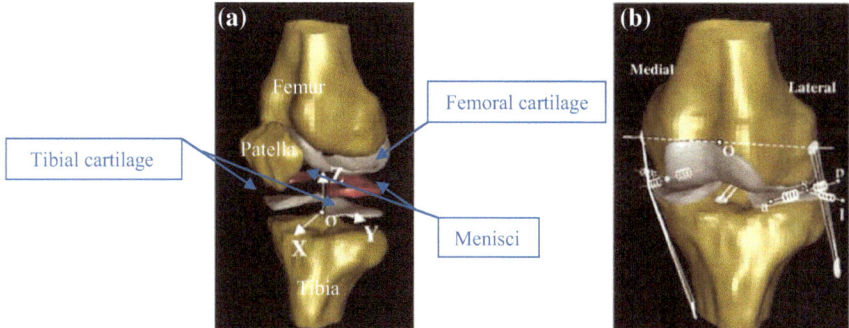

Fig. 1.2 The computation knee joint model developed by Li et al. (1999a): **a** The knee model, including femur, tibia, patella, cartilage layers and menisci; **b** the tibio-femoral joint model, with menisci simulated as equivalent resistant springs

robotic/universal force-moment sensor (UFS) system (Li et al. 1998; Rudy et al. 1996) and knee kinematic data under anterior-posterior tibial loads (up to 10 N) were obtained. Such kinematic data were then compared with the results (tibia anterior-posterior translation) calculated from the FE solution.

The agreement between the FEA and the experimental results has proven the validity of the computational model that may provide useful information for the development of more advanced computational knee models. Thus, the methodology developed in their study can be a valuable benchmark for further analysis of knee joint function under various loading conditions, such as normal gait, standing up from a chair, stair climbing or a variety of rehabilitation exercises. These models also have the potential to be protracted to simulate the effect of ligament reconstruction.

Later, Li et al. (2001a, b, 2002, 2005) extended the work to study the contact kinematics at the tibio-femoral joint and perform FEA studies on the knee, taking into account cartilage thickness variability and ligament injuries. These studies may provide an accurate method for the investigation of articular contact of the knee and may be useful in the investigation of ligament deformation and articular cartilage damage during in vivo functional loading conditions. Consequently, they provide an important insight as to the mechanisms that contribute to OA progression.

Conclusively, this method is the most popular and has been used extensively through the literature to create 3D knee joint models used in the majority of FEA studies (Bachtar et al. 2006; Beillas et al. 2004; Benzakour et al. 2010; Bideau et al. 2011; Blecha et al. 2005; Büchler et al. 2002; Carter and Wong 2003; Dar and Aspden 2003; DeFrate et al. 2004; Donahue et al. 2002; Donzelli et al. 1999; Duchemin et al. 2008; Hirokawa and Tsuruno 2000; Izaham et al. 2012; Johansson et al. 2000; Morimoto et al. 2009; Peña et al. 2005a, 2006a, b, 2007, 2008; Un 2001; Un and Spilker 2006; Vadher et al. 2006; Yang et al. 2010a, b; Yang 2010a; Yao et al. 2006b; Yucesoy et al. 2002; Zhang et al. 1999a, b; Zheng et al. 2014; Zheng 2014; Zhu et al. 2015; Zielinska and Donahue 2006; Fernandes 2014; Gardiner and Weiss 2003; Gardiner et al. 2001; Ghadiali et al. 2004; Harris et al. 2012; Jolivet et al. 2001; Li et al. 2009; Limbert et al. 2004; Little et al. 1986; Meakin et al. 2003; Pandy et al. 1997; Penrose et al. 2002; Piazza and Delp 2001; Song et al. 2004; Wu and Herzog 2000).

1.2 Material Properties of Hard and Soft Tissues

The selection of material characteristic of FE models to better reflect tissue behavior is one of the most critical yet challenging steps in biomechanical FEA studies, since the material properties of hard and soft tissues remain rather controversial and might vary between in vivo and in vitro (Kluess et al. 2009). That explains the difficulty of testing the material properties of tissues in vivo, leading to such controversy in assigning appropriate material properties in FEA.

Several studies have investigated the role of the mechanical properties of the bones, the articular cartilage and the menisci. The aim is to observe the stress and strain distribution in the healthy knee joint and in the joint following various amounts of HTO, meniscectomy or ligament injuries (Peña et al. 2005a, 2006a; Zheng 2014; Zhu et al. 2015; Zielinska and Donahue 2006; Yang 2010a; Yao et al. 2006b; Wilson et al. 2003).

1.2.1 Material Properties of Articular Cartilage

The variation of cartilage thickness was essential during either manual or automatic segmentation processes of FE modelling, since the articular cartilage is relatively thin in the knee joint. The effect of cartilage thickness variation on contact stress was quantified through five FE models of a knee in the study by Li et al. (2001a, b). One was the average FE model constructed from the mean values of the digitized contours of the cartilage and the other four were constructed by varying the mean value of thickness by ±5% and ±10%. Their results demonstrated that under axial tibial compression, variations of cartilage thickness may result in a difference of approximately 10% in peak contact stresses in the cartilage. Therefore, the variation of cartilage thickness was considered to have significant influence on the FE results. Further investigation by Li et al. (2001a, b). into the effect of variation in material properties of soft tissues on contact stress indicated that the stress distribution is also sensitive to the accuracy of material properties of the cartilage model, especially to the variation in the Poisson's ratio.

In parametric studies by Peña et al. (2006a), an increasing Young's modulus of the cartilage led to a rise in contact pressures, maximum compression and shear. Reduction of Poisson's ratio of the cartilage increased the maximum compression and shear stress, but decreased the contact pressure.

Li et al. (1999a) also showed that decreasing the Poisson's ratio or increasing the Young's modulus may lead to substantial rises in shear stresses inside the cartilage. Compared with von Mises stress and hydrostatic pressure, the surface pressure is less sensitive to material property change.

In their studies, Wilson et al. (2003, 2004) computed the maximum shear stress in an axi-symmetric biphasic FE model of the knee joint that consisted of the femoral condyle, tibial plateau, articular cartilage, menisci and a zone of calcified cartilage. Transversely isotropic poroelastic material properties were assigned to the menisci and cartilage and bones were modeled as being much more stiffly elastic isotropic.

Other previous studies by Un (2001) and Un and Spilker (2006), developed an approximate method for simulating the 3D contact of soft biphasic tissues in a diarthrodial joint under physiological loading. A model was constructed of a cartilage layer on top of rigid bones. A biphasic constitutive relation was used to define the material properties of the articular cartilage.

Herzog (2004) investigated the effect of fluid boundary conditions at the articular surfaces on the contact mechanics, in terms of load-sharing and fluid flow properties using variable permeability (Holmes 1986; Fung 1993). The axi-symmetric FEA model consisted of two spheres with a cartilage layer mounted on a layer of bone. The test conditions included a sealed surface, an open surface and open surfaces with variable permeability on the cartilage. The results showed that the closed surface model failed to predict relaxation times and load-sharing properly, while both open surface models gave good agreement of the relaxation time and load-sharing compared to the experimental data (Herzog et al. 1998; Clark et al. 2003).

Donzelli et al. (1999) developed a transversely isotropic, biphasic model of the cartilage to predict high levels of stress and areas of cartilage failure. An axi-symmetric FEA model was constructed with two uniformly thick cartilage layers to study the effect of transverse isotropy and tissue curvature. The results revealed that the transversely isotropic model predicted peak stresses at the cartilage surface and at the cartilage bone interface that was in agreement with results from impact loading (Armstrong et al. 1985; Atkinson and Haut 1995; Haut et al. 1995; Atkinson et al. 1997; Thompson et al. 1991; Vener et al. 1992; Newberry et al. 1997), while isotropic models only predict peak stresses at the cartilage-bone interface (Donzelli et al. 1997; Ateshian and Wang 1995; Ateshian et al. 1994).

Other previous studies addressed articular cartilage as a composite material with highly anisotropic material properties (Laasanen et al. 2003). However, under dynamic loading, the mechanical response of cartilage can be modeled as an isotropic material. A majority of previously published studies have modeled cartilage as a linear elastic isotropic material, because the loading time of interest corresponding to that of a fully extended leg touching the ground is far less than the viscoelastic time constant of cartilage (1500 s) (Armstrong et al. 1984). Moreover, this is considered appropriate due to the elastic response of cartilage during activities involving loading frequencies greater than 1 Hz, such as walking or stair climbing (Besier et al. 2005). This is considered sufficiently accurate to predict instantaneous cartilage response, as demonstrated by Donzelli et al. (1999), who proved that there were no significant changes in the cartilage contact responses shortly after loading.

Blankenvoort et al. (Blankevoort and Huiskes 1996a, b; Blankevoort et al. 1991) and Mommersteeg et al. (1996) modeled the cartilage as elastic isotropic with Young's modulus of 5 MPa and Poisson's ratio of 0.45. This was chosen because the instantaneous response of cartilage to loading corresponds to Young's modulus of 5–15 MPa and a 0.5 Poisson's ratio and the response to long-term loading corresponds to Young's modulus close to 1 MPa and Poisson's ratio ranging from 0 to 0.4 (Kempson 1980).

Subsequently, numerous 3D FEA knee investigations (Li et al. 1999a, 2001a, b; Peña et al. 2006a, b, 2008; Li et al. 2002) have based their properties on the studies of Blankevoort et al. (Blankevoort and Huiskes 1996a, b; Blankevoort et al. 1991).

Donahue et al. (2002, 2003) and Zielinska and Donahue (2006) also modeled the cartilage as isotropic elastic based on the loading time of interest and the stress relaxation of 1500 s (Keer et al. 1990; Armstrong et al. 1984). The Young's modulus was 15 MPa and Poisson's ratio was 0.475, based on experimental measurements from Shepherd and Seedham (1999).

Nevertheless, taking into account a linear elastic behavior law has its limitations. Indeed, during large deformations, the elastic material behavior is not linear. This is the reason why it is more appropriate to consider cartilage as a hyperelastic material. There are several hyper-elasticity models of different complexities. For the modeling of cartilage behavior in the field of large deformations, the Mooney-Rivlin model (Mooney 1940; Rivlin 1948) is widely used in literature (Namani et al. 2003; Anderson et al. 2005).

The strain energy density function for an incompressible Mooney-Rivlin material is:

$$W = C_1 (\overline{I_1} - 3) + C_2 (\overline{I_1} - 3), \tag{1.1}$$

where C_1 and C_2 are empirically determined material constants (coefficients) and I_1 and I_2 are the first and second invariant of the unimodular component of the left Cauchy-Green deformation tensor:

$$\overline{I_1} = J^{-2/3} * I_1, \tag{1.2}$$

$$I_1 = \lambda_1^2 + \lambda_2^2 + \lambda_3^2, \tag{1.3}$$

$$J = \det(F) = \lambda_1 \lambda_2 \lambda_3, \tag{1.4}$$

$$\overline{I_2} = J^{-4/3} * I_2, \tag{1.5}$$

$$I_2 = \lambda_1^2 \lambda_2^2 + \lambda_2^2 \lambda_3^2 + \lambda_3^2 \lambda_1^2, \tag{1.6}$$

where F is the deformation gradient. For an incompressible material, $J = 1$.

Generally speaking, the simplest existing model is the Neo-Hookean hyperelastic model, which can adequately describe large strains and deformations (Bathe 1996). It is a particular case of a Mooney-Rivlin hyperelastic model and it was widely used in the literature (Anderson et al. 1993, 2005; Büchler et al. 2002; Hodge et al. 1986) in order to describe the mechanical behavior of cartilage. This neo-Hookean behavior is defined by a shear modulus (G) and a bulk modulus (K), which depend on the Young's modulus (E) and the Poisson's ratio (υ).

The shear modulus (G) and the bulk modulus (K) are defined in the following forms:

$$G = \frac{E}{4(1 + \upsilon)}, \tag{1.7}$$

$$K = \frac{E}{3(1 - 2\upsilon)}. \tag{1.8}$$

In a few FEA studies (Setton et al. 1994; Wu and Herzog 2000; Wu et al. 1997; Zhang et al. 1999a, b; Athanasiou et al. 1991), cartilage was modeled as a poroelastic material. As an example, in the study by Zhang et al. (1999a, b), a 3D FE model of the human knee joint was created in order to study the interaction between fluid and the solid matrix of the meniscus under kinematic loading conditions. The cartilage and menisci were modeled as isotropic poroelastic materials with a Young's modulus (E) of 0.7 MPa, a Poisson's ratio (υ) of 0.1 and a Permeability coefficient (k) of 2.1710^{-15} m^4/Ns.

Finally, some previous works (Shirazi and Shirazi-Adl 2009a, b; Shirazi et al. 2008; Adouni et al. 2012) used depth-dependent isotropic hyperelastic (Ogden-Compressible) material properties for the non-fibrillar solid matrix of cartilage layers, with the elastic modulus varying linearly from 10 MPa at the surface to 18 MPa at the deep zone and a Poisson's ratio of 0.49. In the short-term stance phase loading, this model with a compressible material was initially verified, through additional simulations, to be equivalent to the incompressible elastic model with much smaller moduli (within the range of 0.3–1.2 MPa).

Following a different path, a few studies (Halonen et al. 2014; Julkunen et al. 2007; Mow and Guo 2002; Wilson et al. 2003; Li and Herzog 2004; Li et al. 1999b, 2000, 2005, 2009) defined cartilage as a fibril-reinforced poroviscoelastic material, in order to investigate the effects of collagen fibril network stiffness, permeability, nonfibrillar matrix modulus and fluid flow boundary conditions on the creep response in cartilage.

In summary, the material properties of the articular cartilages evoked in the literature are shown in Tables 1.1 and 1.2.

1.2.2 Material Properties of Menisci

The knee joint also contains two menisci, comprised of both a medial and a lateral component situated between the corresponding femoral condyle and attached to the tibial plateau by the coronary ligaments (Makris et al. 2011; Kohn and Moreno 1995). Each is a glossy white, complex tissue composed of fibers running in the circumferential direction. Each region has a specific innervation and vascularization (McDermott et al. 2004, 2008; Kohn and Moreno 1995; Kusayama et al. 1994; Ghadially et al. 1978; Shaffer et al. 2000; Greis et al. 2002; Makris et al. 2011). Both menisci are critical components of a healthy knee joint, and its modeling has varied in different knee models.

For example, Li et al. (1999a, 2001a, b, 2002) employed nonlinear spring elements to simulate the equivalent resistance of the menisci.

In other studies developed by Trad et al. (2017), Peña et al. (2005a, 2006a, b, 2008) and LeRoux and Setton (2002), the menisci were modeled as a linear elastic isotropic material with Young's modulus E = 59 MPa and Poisson's ratio υ = 0.49.

In other previous studies of contact analysis (Donahue et al. 2002, 2003; Fithian et al. 1990; Guess et al. 2010; Lechner et al. 2000; Mononen et al. 2012; Sweigart

Table 1.1 Synthesis of the behavioral laws defined in the literature to model the mechanical behavior of cartilaginous tissue

References	Behavioral laws	Material properties	
		E	υ
Cohen et al. (2003), Bei et al. (2004), Bei and Fregly (2004), Blankevoort et al. (1991)	Linear elastic isotropic	4	0.45
Blankevoort and Huiskes (1996), Blankevoort et al. (1991), Kempson (1980), Li et al. (2001a, b, 2002), Mommersteeg et al. (1996), Mow et al. (1982)	Linear elastic isotropic	5	0.45
Agneskirchner et al. (2004), Athanasiou et al. (1995), Carter and Wong (2003), Fernandes (2014), FuJISAwA et al. (1979), Peña et al. (2005a, 2006b, 2007), Yao et al. (2006a), Wan et al. (2013), Zheng et al. (2014), Trad et al. (2017)	Linear elastic isotropic	5	0.46
Peña et al. (2006b, 2008), Li et al. (1999b), Donzelli et al. (1999)	Linear elastic isotropic	9	0.46
Li et al. (2001a, b), Setton et al. (1993), Wang et al. (2014), Shriram et al. (2017), Richard et al. (2013), Łuczkiewicz et al. (2016)	Linear elastic isotropic	10	0.4
Kempson (1979), Brown and DiGioia (1984)	Linear elastic isotropic	10.35	0.4
Russell et al. (2006), von Eisenhart-Rothe et al. (1997), Mavčič et al. (2000), Hodge et al. (1986)	Linear elastic isotropic	12	0.42
Bendjaballah et al. (1995), Brown et al. (1983), Hayes et al. (1972), Hayes and Mockros (1971), Moglo and Shirazi-Adl (2003), Ramaniraka et al. (2005a, b, 2007)	Linear elastic isotropic	12	0.45
Anderson et al. (1993), Rapperport et al. (1985), Wei et al. (2005), Yang and Radin (1990), Zaki et al. (2002), Goto et al. (2002), Nakajima et al. (1994)	Linear elastic isotropic	15	0.45
Oloyede et al. (1992), Repo and Finlay (1977), Beillas et al. (2004)	Linear elastic isotropic	20	0.45
Bachtar et al. (2006), Dalstra et al. (1995)	Linear elastic isotropic	25	0.3

(continued)

Table 1.1 (continued)

References	Behavioral laws	Material properties	
		E	υ
Zhang et al. (1999a, b)	Poroelastic isotropic	0.7 (k = 2.17*10⁻¹⁵ m⁴/Ns)	0.1
Shirazi and Shirazi-Adl (2009a, b), Shirazi et al. (2008), Adouni et al. (2012)	Hyperelastic isotropic	10–18	0.49
Anderson et al. (2008), Harris et al. (2012)	Isotropic, nearly incompressible, hyperelastic Neo-Hookean	G = 13.6 k = 1359	0.495
Zahnert et al. (2000), Ghadially et al. (1978)	Hyperelastic Neo-Hookean	E = 3.4	0.49
Benvenuti (1998), Büchler et al. (2002), Kempson (1979), Kempson et al. (1976)	Hyperelastic Neo-Hookean	E = 10	0.4
Anderson et al. (2005)	Hyperelastic Mooney-Rivlin	C_1 = 4.1 C_2 = 0.41	0.4
Namani et al. (2003)	Hyperelastic Mooney-Rivlin	$0.2 \leq C_1 + C_2 \leq 2$	0.125

E: Young's modulus (MPa), υ: Poisson's ratio

Table 1.2 Poroviscoelastic material properties implemented for articular cartilages

Material properties	Femoral cartilage	Tibial cartilage	References
E_0 (MPa)	0.92	0.18	Halonen et al. (2014), Julkunen et al. (2007)
E_ε (MPa)	150	23.6	Halonen et al. (2014), Julkunen et al. (2007)
E_m (MPa)	0.215	0.106	Halonen et al. (2014), Julkunen et al. (2007)
η (MPa s)	1062	1062	Julkunen et al. (2007), Halonen et al. (2014)
K_0 (10⁻¹⁵ m⁴/Ns)	6	18	Julkunen et al. (2007), Halonen et al. (2014)
$υ_m$	0.15	0.15	Halonen et al. (2014), Wilson et al. (2003)
M	5.09	15.64	Julkunen et al. (2007), Halonen et al. (2014)
n_f	0.8–0.15z	0.8–0.15z	Halonen et al. (2014), Mow and Guo (2002)

E_0: initial fibril network modulus, E_ε: strain-dependent fibril network modulus, η: damping coefficient, E_m: nonfibrillar matrix modulus, k_0: initial permeability, $υ_m$: Poisson's ratio, M: exponential term for the strain dependent permeability, n_f: fluid fraction, and z: normalized depth

et al. 2004; Tissakht and Ahmed 1995; Vaziri et al. 2008; Yang et al. 2010a, b; Yao et al. 2006b; Zielinska and Donahue 2006; Netravali et al. 2011; Wang et al. 2014), the material properties of menisci were defined using linearly elastic transversely isotropic materials with a Young's modulus of 150 MPa in the circumferential direction and 20 MPa in the axial and radial directions. The Poisson's ratio was 0.2 in the circumferential and radial directions and 0.3 in the axial direction.

Based on the studies by Donahue et al. (2002) and Zielinska and Daunahue (2006), the menisci were attached to the tibial plateau at the meniscal horns using a set of linear springs. At each horn attachment, ten linear springs with a stiffness of 200 N/mm attached the horn to the tibial plateau for a total horn stiffness of 2000 N/mm. A transverse ligament with a stiffness of 900 N/mm was modeled as a linear spring and was attached to the anterior horns of the lateral and medial menisci. This method has been used widely for attaching the menisci to the tibia in 3D FEA (Peña et al. 2005a, 2006a, b, 2008; Zielinska and Donahue 2006; Donahue et al. 2002, 2003).

Based on the study by Athanasiou et al. (1991), numerous studies (Korhonen et al. 2003; Li et al. 1999b, 2001a, b, 2005; Wilson et al. 2003; Zhang et al. 1999a, b; Mow and Guo 2002; Julkunen et al. 2007; Halonen et al. 2013, 2014; Dabiri and Li 2013; Bendjaballah et al. 1995; Makris et al. 2011) have described the menisci as cartilaginous tissues displaying poroelastic behavior. At that time, the menisci were assumed to be fibril-reinforced fluid-saturated materials. A fibril-reinforced constitutive law was used that models the solid of the tissue as a linear non-fibrillar matrix reinforced by a nonlinear fibrillar matrix (Li et al. 1999b).

As an example, the poroelastic isotropic behavior was used in the study by Zhang et al. (1999a, b) and Bendjaballah et al. (1995). In their study, the menisci were modeled as a poroelastic isotropic material ($E = 20$ MPa, $\upsilon = 0.1$ and $k = 1.2610^{-15}$ m^4/Ns) reinforced with collagen fibers.

Finally, in the studies by Zheng (2014), Adouni et al. (2012) and Mesfar and Shirazi-Adl (2005), a compressible hyperelastic model was employed for the menisci with a Young's modulus of 10 MPa and a Poisson's ratio of 0.45. Moreover, Halloran et al. (2010) used an ogden material model with material properties based on heel pad indentation tests by Erdemir et al. (2006, 2007, 2009).

In summary, the material properties of the menisci evoked in literature are shown in Tables 1.3 and 1.4.

1.2.3   Material Properties of Ligaments

Other important structures that ensure knee joint compliance and stability required for optimal daily functions are the ligaments. The major ligaments include the Medial Collateral Ligament (MCL), deep (capsular) fibers of MCL (CMCL), Lateral Collateral Ligament (LCL), Anterior Cruciate Ligament (ACL) and Posterior Cruciate Ligament (PCL). This ligamentous structure connects the knee joint and attaches the bones to other bones. Due to the relative incongruence of the articular surfaces, ligaments play a key role in providing passive stability to the joint throughout its whole range of motion and allow for complex mechanical responses to different types of physiological load. A full understanding of the role of each individual ligament in the restraining of motion is essential for the development of an adequate diagnostic and assessment of surgical procedures (Peña et al. 2006a; Blankevoort and Huiskes 1991). Hence, it is important to predict the

Table 1.3 A summary of the material properties assigned to the menisci

References	Behavior laws	Material properties	
		Young's modulus (E) (MPa)	Poisson's ratio (υ)
Bendjaballah et al. (1995), Peña et al. (2008), Spilker et al. (1992), Wilson et al. (2003)	Linear elatistic isotropic	20	0.45
Peña et al. (2007)	Linear elatistic isotropic	20	0.49
Trad et al. (2017), Zheng (2014), Zheng et al. (2014), Wan et al. (2013), Peña et al. (2005a, 2006a, b), LeRoux and Setton (2002), Fernandes (2014)	Linear elatistic isotropic	59	0.49
Fithian et al. (1990), Beillas et al. (2004)	Linear elatistic isotropic	250	0.45
Aspden (1985), Donahue et al. (2002), Fithian et al. (1989), Hefzy et al. (1987), Perie and Hobatho (1998), Schreppers et al. (1990), Skaggs et al. (1994), Tissakht and Ahmed (1995), Whipple et al. (1984)	Linear elatistic transversally isotropic	$E_c = 140$ $E_a = E_r = 20$	$\upsilon_c = \upsilon_r = 0.2$ $\upsilon_a = 0.3$
Fithian et al. (1990), Guess et al. (2010), Lechner et al. (2000), Mononen et al. (2012), Netravali et al. (2011), Sweigart et al. (2004), Tissakht and Ahmed (1995), Vaziri et al. (2008), Wang et al. (2014), Yao et al. (2006b), Zielinska and Donahue (2006)	Linear elatistic transversally isotropic	$E_c = 150$ $E_a = E_r = 20$	$\upsilon_c = \upsilon_r = 0.2$ $\upsilon_a = 0.3$
Łuczkiewicz et al. (2016), Yang et al. (2010a, b), Yang (2010a)	Linear elatistic transversally isotropic	$E_r = 120$ $E_a = E_c = 20$ $G_{ra} = 8.33$ $G_{rc} = G_{ac} = 57.7$	$\upsilon_{rc} = \upsilon_{ac} = 0.3$ $\upsilon_{ra} = 0.2$
Adouni et al. (2012), Zheng et al. (2014), Mesfar and Shirazi-Adl (2005)	Hyperelastic	10	0.45
Li et al. (1999a, 2001a, b, 2002)	Nonlinear spring elements	–	–

G: shear modulus, subscripts a, c, r refer to the constants in the axial, circumferential and radial directions, respectively

Table 1.4 Poroelastic material properties implemented for the menisci

Material properties	Menisci	References
E_0 (MPa)	28	Dabiri and Li (2013), Halonen et al. (2014), Li et al. (2000, 2009), Gu and Li (2011)
E_m (MPa)	0.5	Dabiri and Li (2013), Halonen et al. (2014), Li et al. (2000, 2009), Gu and Li (2011)
K_0 (10^{-15}m^4/Ns)	1.25	Makris et al. (2011), Halonen et al. (2014)
υ_m	0.36	Dabiri and Li (2013), Halonen et al. (2014), Li et al. (2000, 2009), Gu and Li (2011)
M	5.09	Halonen et al. (2014)
n_f	0.72	Makris et al. (2011), Halonen et al. (2014)

mechanical behavior of these biological tissues and provide information that would otherwise be difficult to obtain from experiments.

Due to the intricate anatomy and structural diversity of the knee joint, some FEA studies represented the knee ligaments with a one-dimensional (1D) truss or beam (Adouni et al. 2012; Bendjaballah et al. 1997; Penrose et al. 2002) or spring elements (Beillas et al. 2004; Donahue et al. 2002; Li et al. 1999a, 2002; Moglo and Shirazi-Adl 2003) with simplified material properties.

Based on the study by Mesfar and Shirazi-Adl (2005), ligaments were individually modeled, in the study by Adouni et al. (2012), by a number of uniaxial elements with different initial pre-strains, nonlinear (tension-only) material properties, and initial cross-sectional areas of 42, 60, 18 and 25 mm^2 for ACL, PCL, LCL and MCL, respectively. Even with the increased simplicity of obtaining kinematic and force data, this approach does not permit determination of the stress distribution in the ligaments.

The challenge when using 3D representations of the ligaments is to develop a material model that can perfectly distinguish the nonlinear behavior of ligaments.

In regard to this matter, the vast majority of contact analyses studies (Blankevoort and Huiskes 1996a, b; Blankevoort et al. 1991; Donahue et al. 2003; Li et al. 1999a, 2001a, b; Mommersteeg et al. 1996; Peña et al. 2005a, 2006a, b, 2008; Yao et al. 2006b; Zielinska and Donahue 2006; Yang et al. 2010a, b; Yang 2010a; Yang and Radin 1990; Donahue et al. 2002) have modeled the ligaments as nonlinear spring elements. A piecewise function developed by Blankevoort et al. (Blankevoort and Huiskes 1991; Blankevoort et al. 1991) defines the behavior of the ligaments at high strains (>6%) and low strains (<6%). The material properties of the different ligaments were defined using optimization between cadaver experiments and numerical simulations. The piecewise function describing the force-displacement relationship of a non-linear ligament spring element was defined as follows:

$$f = \begin{cases} \frac{1}{4}\kappa\varepsilon^2/\varepsilon, & 0 \le \varepsilon \le 2\varepsilon_1 \\ \kappa(\varepsilon - \varepsilon_1), & \varepsilon > 2\varepsilon_1 \\ 0, & \varepsilon < 0 \end{cases} \quad , \tag{1.9}$$

where f is the tensile force; κ is the ligament stiffness parameter used by Blankevoort et al. (Blankevoort and Huiskes 1991, 1996a, b; Blankevoort et al. 1991) and obtained from experimental data available in the literature (Andriacchi et al. 1983; Butler et al. 1986; Wismans et al. 1980; Danylchuk et al. 1978); ε_1 is the non-linear strain parameter, assumed to be 0.03 (Butler et al. 1986; Blankevoort and Huiskes 1991; Blankevoort et al. 1991); and ε is the strain in the ligaments (Blankevoort and Huiskes 1991; Blankevoort et al. 1991) defined as

$$\varepsilon = (L - L_0)/L_0, \tag{1.10}$$

where L is the ligament length after deformation and L_0 is the reference length of the ligament at which the ligament starts to bear the tensile force (Blankevoort and Huiskes 1991; Blankevoort et al. 1991). It was derived from the initial strain (ε_r) and the initial length of the ligament (L_r) as follows:

$$L_0 = L_r/(\varepsilon_r + 1), \tag{1.11}$$

$$\varepsilon_r = (L_r - L_0)/L_0, \tag{1.12}$$

where L_r is determined from the length of the ligament in MR images. The initial strain (ε_r) was found by minimizing the difference in knee kinematics between experimental data and a 3D knee model created by Blankevoort et al. (Blankevoort and Huiskes 1991; Blankevoort et al. 1991).

The ligaments were modeled according to their functional bundles based on actual ligament anatomy. Each functional bundle of the ligaments was depicted with a non-linear spring element. The stiffness parameter (κ) and the initial strain (ε_r) of the different ligament bundles reported in several previous studies are listed in Table 1.5. Positive values of (ε_r) correspond to initial tension, while negative values correspond to initially slack ligament bundles. The slack length of a ligament element is defined as the difference between (L_0) and the length of the ligament element measured at full extension when the knee specimen was scanned in the MRI system.

Various articles in the literature (Dhaher et al. 2010; Ellis et al. 2006; Fernandes 2014; Gardiner and Weiss 2003; Gardiner et al. 2001; Limbert et al. 2004; Peña et al. 2005b, 2006a; Weiss and Gardiner 2001; Weiss et al. 1995, 1996, 2002 Woo et al. 1999; Yagi et al. 2002; Yamamoto et al. 1998; Zheng 2014; Pioletti et al. 1998; Pioletti and Rakotomanana 2000; Ramaniraka et al. 2005b) have modeled the ligaments as transversely isotropic hyperelastic material when specifically investigating the ligaments' behavior at loading, since the hyperelastic behavior has the capability of providing more accurate stress prediction under moderate loading, which is often a more clinically realistic case.

Table 1.5 Non-linear spring element parameters assigned to the ligaments

Knee ligaments	Ligament bundles	κ (N)	ε_r	Slack length (mm)
LCL	Anterior	2000	−0.25	−0.04
	Posterior	2000	−0.05	−0.04
	Superior	2000	0.08	−0.04
MCL	Anterior	2750	0.04	0.2
	Inferior	2750	0.04	0.4
	Superior	2750	0.03	0.2
ACL	Anterior	5000	0.06	0.8
	Posterior	5000	0.1	0.3
PCL	Anterior	9000	−0.24	2.3
	Posterior	9000	−0.03	3
CMCL	Anterior	1000	–	1.1
	Posterior	1000	–	0.3

For example, in the study by Zheng (2014), the hyperelasticity of a ligament was represented by an incompressible Neo-Hookean model with parameters C_{10} and D_1 for ligaments, in which the C_{10} refers to the stiffness of the Neo-Hookean strain energy function and D_1 denotes the inverse of the bulk modulus. These parameters were taken from the studies by Biscoff et al. (2008). Moreover, the relative Neo-Hookean constants of specific ligaments were calculated according to the experimental data from the study presented by Arnoux et al. (2002). All of the constants of the ligaments calculated are summarised in Table 1.6.

Other studies reported by (Łuczkiewicz et al. 2016; Gardiner and Weiss 2003; Peña et al. 2006a) used a nearly incompressible, hyperelastic, transversely isotropic material model targeted at describing the behavior of ligaments. The strain energy function (Eq. 1.13) contains a Neo-Hookean form with two constants C_1 and D_1 (Table 1.7) defining the matrix substance and the transverse part $F_2(\lambda)$ depending on the stiffness of the collagen fibres (Holzapfel 2002; Weiss et al. 1996):

$$\Phi = C_1 \left(\overline{I_1} - 3 \right) + \frac{1}{D_1} (J - 1)^2 + F_2(\lambda), \tag{1.13}$$

$$J = Det\,F, \tag{1.14}$$

$$\overline{I_1} = tr\,\overline{B}. \tag{1.15}$$

Table 1.6 Material properties of the ligaments used in the FEA model (Zheng 2014)

Knee ligaments	C_{10}	D_1
ACL	1.95	0.00683
PCL	3.24	0.0042
LCL	1.45	0.00127
MCL	1.45	0.00127

Table 1.7 Hyperelasticity material parameters for ligaments (Łuczkiewicz et al. 2016; Gardiner and Weiss 2003; Peña et al. 2006a)

	C_1 (MPa)	D_1 (MPa^{-1})	C_3 (MPa)	C_4	C_5 (MPa)	λ^*
ACL	1.95	0.01366	0.0139	116.22	535.039	1.046
PCL	3.25	0.0082	0.1196	87.178	431.036	1.035
LCL	1.44	0.00252	0.57	48	467.1	1.062
MCL	1.44	0.00252	0.57	48	467.1	1.062

The function $F_2(\lambda)$ fulfils the following conditions:

$$\lambda \frac{\partial F_2(\lambda)}{\partial \lambda} = \begin{cases} 0, & \lambda \le 1 \\ C_3(\exp(C_4(\lambda - 1)) - 1), & 1 < \lambda < \lambda^*, \\ C_5\lambda + C_6, & \lambda < \lambda^* \end{cases} \qquad (1.16)$$

where λ is the fiber stretch, λ^* is the limit value of the stretch corresponding to the straightened fiber, C_3, C_4 and C_5 are the material constants given in Table 1.7, and the constant C_6 was calculated from the continuity condition for λ^*.

The following relation relates the fiber stretch with the deformation gradient F:

$$\lambda \mathbf{a}(\mathbf{x}) = \mathbf{F}(\mathbf{X}).\mathbf{a}_0(\mathbf{X}), \qquad (1.17)$$

where a and $\mathbf{a}_0$ are the orientations of fibres in the current configuration $\mathbf{x}$ and the reference configuration $\mathbf{X}$, respectively.

Finally, a few studies in the literature (Kiapour et al. 2014; Westermann et al. 2013) adopted an incompressible anisotropic hyperelastic model using the Holzapfel-Gasser-Ogden (HGO) material model (Gasser et al. 2006).

In fact, the HGO model is a hyperelastic, anisotropic material model that was evolved to model the criss-crossed fibrous soft tissues (Girgis et al. 1975). As a brief definition, an isotropic non-collagenous ground matrix was modeled by the incompressible hyperelastic neo-Hookean component of the strain energy density (SED) function, whereas the transversely isotropic fibrous component was modeled following this function (Eq. 1.18) developed by Gasser et al. (2006):

$$\overline{\Psi}(\overline{C}, H_i) = \overline{\Psi}_g(\overline{C}) + \sum_{i=1}^{2} \Psi_{fi}(\overline{C}, H_i(\mathbf{a}_{0i}, \kappa)), \qquad (1.18)$$

where $\overline{\Psi}_g$ and Ψ_{fi} are the respective isotropic and anisotropic components of the SED, $\mathbf{a}_0$ is the mean orientation of the fibers, $H(\mathbf{a}_0, \kappa)$ is the structure tensor, and κ is the dispersion parameter for the fiber family.

1.2.4 Material Properties of Bony Structure

Since the stiffness of bone is much higher than that of the relevant soft tissues, its influence in many studies available in the literature was minimal (Donahue et al. 2002; Duchemin et al. 2008; Fithian et al. 1990; Wei et al. 2005; Viswanath et al. 2007; Peña et al. 2006a). The FEA results are less sensitive to the material properties of a hard bony structure (femur, tibia, patella, fibula).

In order to avoid unnecessary computational costs, previous 3D FEA studies (Abdel-Rahman and Hefzy 1998; Adams et al. 2007; Adouni et al. 2012; Blankevoort et al. 1991; Cowin 2001; Debski et al. 2005; Donahue et al. 2002; Drury et al. 2010, 2011; Ellis et al. 2007; Gatti et al. 2010; Kahle et al. 1998; LeRoux and Setton 2002; Li et al. 1999a; Łuczkiewicz et al. 2016; Luo et al. 1998; Martin et al. 1998; Moore et al. 2010; Peña et al. 2005a, 2006a, b, 2007, 2008; Perie and Hobatho 1998; Piazza and Delp 2001; Terrier et al. 2007; Walia et al. 2013; Wismans et al. 1980; Yang et al. 2010a, b; Yang 2010a; Fernandes 2014) have assumed that the femur and tibia were normally considered as rigid bodies because of their relatively high density and Young's modules compared to the cartilage and menisci in the knee joint.

On the other hand, several articles in the literature considered the bones to be deformable bodies (Chao 2003; Cohen et al. 2001, 2003; Donahue et al. 2002, 2003; Elias et al. 2004; Guess et al. 2010; Kwak et al. 2000; Zheng et al. 2014; Zheng 2014; Duda et al. 1998; Lengsfeld et al. 1998; LeRoux and Setton 2002; Miyoshi et al. 2002a; Yamamoto et al. 1998; Harris et al. 2012; Anderson et al. 2005, 2008; Dabiri and Li 2013; Benli et al. 2008; Izaham et al. 2012; Dalstra et al. 1993, 1995; Dalstra and Huiskes 1995).

For example, Guess et al. (2010) defined the femur and tibia as an isotropic material with a Young's modulus of 20 GPa, a Poisson's ratio of 0.2, and a density of 1600 kg/m^3, while other studies (Zysset et al. 1999; Rho et al. 1993; Wang et al. 2014) considered the cortical bone as an isotropic material with a Young's modulus of 20 GPa, but a Poisson's ratio of 0.3.

In the study conducted by Zheng (2014), the femur, tibia and patella were assumed to be a linear, elastic and isotropic material with Young's modulus of 8 GPa and a Poisson's ratio of 0.3. Another study by Miyoshi et al. (2002a) evaluated the change of stress and displacement with differing shapes of the tibial component in total knee arthroplasty. In fact, the cortical bone was assumed to have a Poisson's ratio of 0.3 and an elastic modulus of 8 GPa. The cancellous bone was assumed to have a Poisson's ratio of 0.2 and an elastic modulus of 1500 MPa.

In the study presented by Donahue et al. (2002), a 3D FE knee joint model was generated based on CT images to determine whether the assumption of bone as a rigid body could affect contact behavior. Under the application of an 800 N compressive load, the maximum pressure, mean pressure, contact area, total contact force and coordinate of the center of pressure were determined. The results revealed that the contact responses within the knee joint have no significant difference whether the bone is assumed to be a rigid body or a deformable body. This explains why bones are considered either as a rigid body or as a linearly elastic isotropic material in the literature.

Ultimately, the material properties of the bone mentioned in the literature are summarized in Tables 1.8 and 1.9.

In summary, the FEA results are considered sensitive to both cartilage thickness and the material properties of hard and soft tissues in the knee joint. Hence, selection of a proper material model and accurate segmentation of the knee joint components are critical in FEA studies.

Table 1.8 A summary of the material properties assigned to the bony structure

References	Behavior law	Material properties			
		Cortical bone		Cancellous bone	
		E	υ	E	υ
Dabiri and Li (2013), Gu and Li (2011)	Linear elastic isotropic	5000	0.3	–	–
Donahue et al. (2002, 2003), Zheng et al. (2014), Zheng (2014)	Linear elastic isotropic	8000	0.3	–	–
Anderson et al. (2008), Harris et al. (2012)	Linear elastic isotropic	17,000	0.29	–	–
Adams et al. (2007), Terrier et al. (2007)	Linear elastic isotropic	17,000	0.3	–	–
Katsamanis and Raftopoulos (1990)	Linear elastic isotropic	19,900	0.36	–	–
Benli et al. (2008), Izaham et al. (2012)	Linear elastic isotropic	20,000	0.3	–	–
Guess et al. (2010)	Linear elastic isotropic	20,000	0.2	–	–
Russell et al. (2006)	Linear elastic homogenous	2000	0.3	120	0.3
Brown and Ferguson (1980), Reilly et al. (1974), Brown et al. (1983)	Linear elastic isotropic	6900	0.26	690	0.3
Miyoshi et al. (2002a, b)	Linear elastic isotropic	8000	0.3	1500	0.2
Lengsfeld et al. (1998)	Linear elastic homogenous	15,000	–	1100	-
Bachtar et al. (2006)	Linear elastic isotropic	17,000	0.3	2000	0.2
Dalstra et al. (1993, 1995), Dalstra and Huiskes (1995)	Linear elastic isotropic	17,000	0.3	800	0.2
Duda et al. (1998)	Linear elastic homogenous	17,000	0.33	15,000	0.3
Büchler et al. (2002), Hayes and Bouxsein (1991), Rice et al. (1988)	Linear elastic non-homogenous	15,000	0.3	–	–

Table 1.9 A summary of previous studies that take into account the subchondral bone

References	Behavior law	Cortical bone		Cancellous bone		Subchondral bone		
		E (MPa)	υ	E (MPa)	υ	Behavior law	E (MPa)	υ
Oonishi et al. (1983)	Linear isotropic heterogeneous	15,000	0.3	1000	0.2	Linear isotropic homogeneous	3000	0.2
Dalstra and Huiskes (1995)	Linear isotropic heterogeneous	17,000	0.3	2000	0.3	Linear isotropic homogeneous	70	0.2
Dalstra and Huiskes (1995)	Linear anisotropic heterogeneous	17,000	0.3	$186 \leq E \leq 2155$	0.3	Linear anisotropic heterogeneous	$1 \leq E \leq 132$	0.2
Besnault (1999)	Linear isotropic heterogeneous	17,000	0.3	2000	0.2	Linear isotropic homogeneous	70	0.2
Wei et al. (2005)	Linear isotropic heterogeneous	17,000	0.28	1100	0.3	Linear isotropic homogeneous	600	0.3
Wei et al. (2005)	Linear isotropic heterogeneous	17,000	0.3	700	0.3	–	–	–

References

Abdel-Rahman, E., & Hefzy, M. S. (1993). A two-dimensional dynamic anatomical model of the human knee joint. *Transactions-American Society of Mechanical Engineers Journal of Biomechanical Engineering, 115,* 357.

Abdel-Rahman, E. M., & Hefzy, M. S. (1998). Three-dimensional dynamic behaviour of the human knee joint under impact loading. *Medical Engineering & Physics, 20*(4), 276–290.

Adams, C. R., et al. (2007). Effects of rotator cuff tears on muscle moment arms: A computational study. *Journal of Biomechanics, 40*(15), 3373–3380.

Adouni, M., Shirazi-Adl, A., & Shirazi, R. (2012). Computational biodynamics of human knee joint in gait: from muscle forces to cartilage stresses. *Journal of Biomechanics, 45*(12), 2149–2156.

Agneskirchner, J. D., et al. (2004). Effect of high tibial flexion osteotomy on cartilage pressure and joint kinematics: A biomechanical study in human cadaveric knees. *Archives of Orthopaedic and Trauma Surgery, 124*(9), 575–584.

Ali, A. A., et al. (2016). Validation of predicted patellofemoral mechanics in a finite element model of the healthy and cruciate-deficient knee. *Journal of Biomechanics, 49*(2), 302–309.

Anderson, D. D., Brown, T. D., & Radin, E. L. (1993). The influence of basal cartilage calcification on dynamic juxtaarticular stress transmission. *Clinical Orthopaedics and Related Research, 286,* 298–307.

Anderson, A. E., et al. (2005). Subject-specific finite element model of the pelvis: development, validation and sensitivity studies. *Journal of Biomechanical Engineering, 127*(3), 364–373.

Anderson, A. E., et al. (2008). Validation of finite element predictions of cartilage contact pressure in the human hip joint. *Journal of Biomechanical Engineering, 130*(5), 51008.

Andriacchi, T. P., et al. (1983). Model studies of the stiffness characteristics of the human knee joint. *Journal of Biomechanics, 16*(1), 23–29.

Armstrong, C. G., Lai, W. M., & Mow, V. C. (1984). An analysis of the unconfined compression of articular cartilage. *Journal of Biomechanical Engineering, 106*(2), 165–173.

Armstrong, C. G., Mow, V. C., & Wirth, C. R. (1985). Biomechanics of impact-induced microdamage to articular cartilage: A possible genesis for chondromalacia patella. In *AAOS Symposium on Sports Medicine: The Knee CV Mosby Co, St. Louis* (pp. 70–84).

Arnoux, P. J., et al. (2002). A visco-hyperelastic model with damage for the knee ligaments under dynamic constraints. *Computer Methods in Biomechanics & Biomedical Engineering, 5*(2), 167–174.

Aspden, R. M. (1985). A model for the function and failure of the meniscus. *Engineering in Medicine, 14*(3), 119–122.

Ateshian, G. A., & Wang, H. (1995). A theoretical solution for the frictionless rolling contact of cylindrical biphasic articular cartilage layers. *Journal of Biomechanics, 28*(11), 1341–1355.

Ateshian, G. A., et al. (1994). An asymptotic solution for the contact of two biphasic cartilage layers. *Journal of Biomechanics, 27*(11), 1347–1360.

Athanasiou, K. A., et al. (1991). Interspecies comparisons of in situ intrinsic mechanical properties of distal femoral cartilage. *Journal of Orthopaedic Research, 9*(3), 330–340.

Athanasiou, K. A., et al. (1995). Effects of excimer laser on healing of articular cartilage in rabbits. *Journal of Orthopaedic Research, 13*(4), 483–494.

Atkinson, P. J., & Haut, R. C. (1995). Subfracture insult to the human cadaver patellofemoral joint produces occult injury. *Journal of Orthopaedic Research, 13*(6), 936–944.

Atkinson, T. S., Haut, R. C., & Altiero, N. J. (1997). Fissuring of articular cartilage during blunt insult: An investigation of failure criteria. *Transactions of Orthopaedic Research Society, 22,* 824.

Bach, B. R., et al. (1992). Force displacement characteristics of the posterior cruciate ligament. *The American journal of sports medicine, 20*(1), 67–72.

Bachtar, F., Chen, X., & Hisada, T. (2006). Finite element contact analysis of the hip joint. *Medical & Biological Engineering & Computing, 44*(8), 643–651.

Bartel, D. L., et al. (1977). Surgical repositioning of the medial collateral ligament. An anatomical and mechanical analysis. *Journal of Bone and Joint Surgery. American Volume, 59*(1), 107–116.

Bathe, K. J. (1996). *Finite element procedures*. Englewood Cliffs, NJ: Prentice-Hall.

Bei, Y., & Fregly, B. J. (2004). Multibody dynamic simulation of knee contact mechanics. *Medical Engineering & Physics, 26*(9), 777–789.

Bei, Y., et al. (2004). The relationship between contact pressure, insert thickness, and mild wear in total knee replacements. *Computer Modeling in Engineering and Sciences, 6,* 145–152.

Beillas, P., et al. (2004). A new method to investigate in vivo knee behavior using a finite element model of the lower limb. *Journal of Biomechanics, 37*(7), 1019–1030.

Bendjaballah, M. Z., Shirazi-Adl, A., & Zukor, D. J. (1995). Biomechanics of the human knee joint in compression: Reconstruction, mesh generation and finite element analysis. *The Knee, 2* (2), 69–79.

Bendjaballah, M. Z., Shirazi-Adl, A., & Zukor, D. J. (1997). Finite element analysis of human knee joint in varus-valgus. *Clinical Biomechanics, 12*(3), 139–148.

Benli, S., et al. (2008). Evaluation of bone plate with low-stiffness material in terms of stress distribution. *Journal of Biomechanics, 41*(15), 3229–3235.

Benvenuti, J. F. (1998). *Modélisation tridimensionnelle du genou humain*. Lausanne: Swiss Federal Institute of Technology.

Benzakour, T., et al. (2010). High tibial osteotomy for medial osteoarthritis of the knee: 15 years follow-up. *International orthopaedics, 34*(2), 209–215.

Besier, T. F., et al. (2005). A modeling framework to estimate patellofemoral joint cartilage stress in vivo. *Medicine and Science in Sports and Exercise, 37*(11), 1924.

Besnault, B. (1999). Modélisation par éléments finis du bassin humain en configuration de chocs automobiles. Ph.D. thesis, ENSAM, Paris.

Beynnon, B., et al. (1996). A sagittal plane model of the knee and cruciate ligaments with application of a sensitivity analysis. *Transactions-American Society of Mechanical Engineers Journal of Biomechanical Engineering, 118,* 227–239.

Bideau, N., et al. (2011). Développement d'un modèle ostéoarticulaire du genou humain pour une analyse dynamique du contact en grands déplacements. In *10e colloque national en calcul des structures* (p. Clé-USB).

Bischoff, J. E., et al. (2008). Advanced material modeling in a virtual biomechanical knee. In *Abaqus Users' Conference*.

Blankevoort, L., & Huiskes, R. (1991). Ligament-bone interaction in a three-dimensional model of the knee. *Journal of Biomechanical Engineering, 113*(3), 263–269.

Blankevoort, L., & Huiskes, R. (1996a). A mechanism for rotation restraints in the knee joint. *Journal of Orthopaedic Research, 14*(4), 676–679.

Blankevoort, L., & Huiskes, R. (1996b). Validation of a three-dimensional model of the knee. *Journal of Biomechanics, 29*(7), 955–961.

Blankevoort, L., Huiskes, R. & De Lange, A. (1988). The envelope of passive knee joint motion. *Journal of Biomechanics*, *21*(9), 705711–709720.

Blankevoort, L., et al. (1991). Articular contact in a three-dimensional model of the knee. *Journal of Biomechanics, 24*(11), 1019–1031.

Blecha, L. D., et al. (2005). How plate positioning impacts the biomechanics of the open wedge tibial osteotomy; a finite element analysis. *Computer Methods in Biomechanics and Biomedical Engineering, 8*(5), 307–313.

Brantigan, O. C., & Voshell, A. F. (1941). The mechanics of the ligaments and menisci of the knee joint. *Journal of Bone and Joint Surgery. American Volume, 23*(1), 44–66.

Brown, T. D., & DiGioia, A. M. (1984). A contact-coupled finite element analysis of the natural adult hip. *Journal of Biomechanics, 17*(6), 437–448.

Brown, T. D., Digioia, A. M., & Mears, D. C. (1983). A contact coupled nonlinear finite element analysis of the hip joint. In *Transactions of 29th Annual Meeting ORS* (p. 66).

Brown, T. D., & Ferguson, A. B. (1980). Mechanical property distributions in the cancellous bone of the human proximal femur. *Acta Orthopaedica Scandinavica, 51*(1–6), 429–437.

Büchler, P., et al. (2002). A finite element model of the shoulder: Application to the comparison of normal and osteoarthritic joints. *Clinical Biomechanics, 17*(9), 630–639.

Butler, D. L., Kay, M. D., & Stouffer, D. C. (1986). Comparison of material properties in fascicle-bone units from human patellar tendon and knee ligaments. *Journal of Biomechanics, 19*(6), 425–432.

Carter, D. R., & Wong, M. (2003). Modelling cartilage mechanobiology. *Philosophical Transactions of the Royal Society of London B: Biological Sciences, 358*(1437), 1461–1471.

Chao, E. Y. S. (2003). Graphic-based musculoskeletal model for biomechanical analyses and animation. *Medical Engineering & Physics, 25*(3), 201–212.

Clark, A. L., et al. (2003). In situ chondrocyte deformation with physiological compression of the feline patellofemoral joint. *Journal of Biomechanics, 36*(4), 553–568.

Cohen, Z. A., et al. (2001). Patellofemoral stresses during open and closed kinetic chain exercises An analysis using computer simulation. *The American Journal of Sports Medicine, 29*(4), 480–487.

Cohen, Z. A., et al. (2003). Computer simulations of patellofemoral joint surgery patient-specific models for tuberosity transfer. *The American Journal of Sports Medicine, 31*(1), 87–98.

Cowin, S. C. (2001). *Bone mechanics handbook*, CRC press.

Crowninshield, R., Pope, M. H., & Johnson, R. J. (1976). An analytical model of the knee. *Journal of Biomechanics, 9*(6), 397–405.

Dabiri, Y., & Li, L. P. (2013). Altered knee joint mechanics in simple compression associated with early cartilage degeneration. *Computational and mathematical methods in medicine.*

Dalstra, M., & Huiskes, R. (1995). Load transfer across the pelvic bone. *Journal of Biomechanics, 28*(6), 715–724.

Dalstra, M., Huiskes, R., & Van Erning, L. (1995). Development and validation of a three-dimensional finite element model of the pelvic bone. *Journal of Biomechanical Engineering, 117*(3), 272–278.

Dalstra, M., et al. (1993). Mechanical and textural properties of pelvic trabecular bone. *Journal of Biomechanics, 26*(4–5), 523–535.

Danylchuk, K. D., Finlay, J. B., & Krcek, J. P. (1978). Microstructural organization of human and bovine cruciate ligaments. *Clinical Orthopaedics and Related Research, 131*, 294–298.

Dar, F. H., & Aspden, R. M. (2003). A finite element model of an idealized diarthrodial joint to investigate the effects of variation in the mechanical properties of the tissues. *Proceedings of the Institution of Mechanical Engineers. Part H, Journal of Engineering in Medicine, 217*(5), 341–348.

Debski, R. E., et al. (2005). Stress and strain in the anterior band of the inferior glenohumeral ligament during a simulated clinical examination. *Journal of Shoulder and Elbow Surgery, 14* (1), S24–S31.

DeFrate, L. E., et al. (2004). In vivo tibiofemoral contact analysis using 3D MRI-based knee models. *Journal of Biomechanics, 37*(10), 1499–1504.

Dhaher, Y. Y., Kwon, T.-H., & Barry, M. (2010). The effect of connective tissue material uncertainties on knee joint mechanics under isolated loading conditions. *Journal of Biomechanics, 43*(16), 3118–3125.

Donahue, T. L. H., et al. (2002). A finite element model of the human knee joint for the study of tibio-femoral contact. *Journal of Biomechanical Engineering, 124*(3), 273–280.

Donahue, T. L. H., et al. (2003). How the stiffness of meniscal attachments and meniscal material properties affect tibio-femoral contact pressure computed using a validated finite element model of the human knee joint. *Journal of Biomechanics, 36*(1), 19–34.

Donzelli, P., et al. (1997). Physiological joint incongruity significantly affects the load partitioning between the solid and fluid phases of articular cartilage. *Transactions of the Orthopedic Research Society, 22, 82.*

Donzelli, P. S., et al. (1999). Contact analysis of biphasic transversely isotropic cartilage layers and correlations with tissue failure. *Journal of Biomechanics, 32*(10), 1037–1047.

Drury, N. J., et al. (2010). The impact of glenoid labrum thickness and modulus on labrum and glenohumeral capsule function. *Journal of Biomechanical Engineering, 132*(12), 121003.

Drury, N. J., et al. (2011). Finding consistent strain distributions in the glenohumeral capsule between two subjects: Implications for development of physical examinations. *Journal of Biomechanics, 44*(4), 607–613.

Duchemin, L., et al. (2008). Prediction of mechanical properties of cortical bone by quantitative computed tomography. *Medical Engineering & Physics, 30*(3), 321–328.

Duda, G. N., et al. (1998). Influence of muscle forces on femoral strain distribution. *Journal of Biomechanics, 31*(9), 841–846.

Elias, J. J., et al. (2004). Evaluation of a computational model used to predict the patellofemoral contact pressure distribution. *Journal of Biomechanics, 37*(3), 295–302.

Ellis, B. J., et al. (2006). Medial collateral ligament insertion site and contact forces in the ACL-deficient knee. *Journal of Orthopaedic Research, 24*(4), 800–810.

Ellis, B. J., et al. (2007). Methodology and sensitivity studies for finite element modeling of the inferior glenohumeral ligament complex. *Journal of Biomechanics, 40*(3), 603–612.

Erdemir, A., et al. (2006). An inverse finite-element model of heel-pad indentation. *Journal of Biomechanics, 39*(7), 1279–1286.

Erdemir, A., et al. (2007). Model-based estimation of muscle forces exerted during movements. *Clinical Biomechanics, 22*(2), 131–154.

Erdemir, A., et al. (2009). An elaborate data set characterizing the mechanical response of the foot. *Journal of Biomechanical Engineering, 131*(9), 94502.

Essinger, J. R., et al. (1989). A mathematical model for the evaluation of the behaviour during flexion of condylar-type knee prostheses. *Journal of Biomechanics, 22*(11–12), 1229–1241.

Fernandes, D. J. C. (2014). Finite element analysis of the ACL-deficient knee. Ph.D. thesis, IST, Universidade de Lisboa, Portugal.

Fithian, D. C., Kelly, M. A., & Mow, V. C. (1990). Material properties and structure-function relationships in the menisci. *Clinical Orthopaedics and Related Research, 252,* 19–31.

Fithian, D. C., et al. (1989). Human meniscus tensile properties: Regional variation and biochemical correlation. *Transactions of the Orthopedic Research Society, 35,* 205.

FuJISAwA, Y., Masuhara, K., & Shiomi, S. (1979). The effect of high tibial osteotomy on osteoarthritis of the knee. An arthroscopic study of 54 knee joints. *The Orthopedic Clinics of North America, 10*(3), 585–608.

Fung, Y.-C. (1993). Mechanical properties and active remodeling of blood vessels. In *Biomechanics* (pp. 321–391). Springer.

Gardiner, J. C., & Weiss, J. A. (2003). Subject-specific finite element analysis of the human medial collateral ligament during valgus knee loading. *Journal of Orthopaedic Research, 21*(6), 1098–1106.

Gardiner, J. C., Weiss, J. A., & Rosenberg, T. D. (2001). Strain in the human medial collateral ligament during valgus loading of the knee. *Clinical Orthopaedics and Related Research, 391,* 266–274.

Garg, A., & Walker, P. S. (1990). Prediction of total knee motion using a three-dimensional computer-graphics model. *Journal of Biomechanics, 23*(1), 45–58.

Gasser, T. C., Ogden, R. W., & Holzapfel, G. A. (2006). Hyperelastic modelling of arterial layers with distributed collagen fibre orientations. *Journal of the Royal Society, Interface, 3*(6), 15–35.

Gatti, C. J., et al. (2010). Development and validation of a finite element model of the superior glenoid labrum. *Annals of Biomedical Engineering, 38*(12), 3766–3776.

Ghadiali, S. N., Banks, J., & Swarts, J. D. (2004). Finite element analysis of active Eustachian tube function. *Journal of Applied Physiology, 97*(2), 648–654.

Ghadially, F. N., et al. (1978). Ultrastructure of rabbit semilunar cartilages. *Journal of Anatomy, 125*(Pt 3), 499.

Gibson, M., et al. (1986). Analysis of the Müller anterolateral femorotibial ligament reconstruction using a computerized knee model. *The American Journal of Sports Medicine, 14*(5), 371–375.

Girgis, F. G., Marshall, J. L., & Jem, A. R. S. A. M. (1975). The cruciate ligaments of the knee joint: Anatomical. functional and experimental analysis. *Clinical Orthopaedics and Related Research, 106,* 216–231.

Goto, K., et al. (2002). Mechanical analysis of the lumbar vertebrae in a three-dimensional finite element method model in which intradiscal pressure in the nucleus pulposus was used to establish the model. *Journal of Orthopaedic Science, 7*(2), 243–246.

Greis, P. E., et al. (2002). Meniscal injury: I. Basic science and evaluation. *Journal of the American Academy of Orthopaedic Surgeons, 10*(3), 168–176.

Gu, K. B., & Li, L. P. (2011). A human knee joint model considering fluid pressure and fiber orientation in cartilages and menisci. *Medical Engineering & Physics, 33*(4), 497–503.

Guess, T. M., et al. (2010). A subject specific multibody model of the knee with menisci. *Medical Engineering & Physics, 32*(5), 505–515.

Halloran, J. P., et al. (2010). Concurrent musculoskeletal dynamics and finite element analysis predicts altered gait patterns to reduce foot tissue loading. *Journal of Biomechanics, 43*(14), 2810–2815.

Halonen, K. S., et al. (2013). Importance of depth-wise distribution of collagen and proteoglycans in articular cartilage—a 3D finite element study of stresses and strains in human knee joint. *Journal of Biomechanics, 46*(6), 1184–1192.

Halonen, K. S., et al. (2014). Deformation of articular cartilage during static loading of a knee joint–experimental and finite element analysis. *Journal of Biomechanics, 47*(10), 2467–2474.

Harris, M. D., et al. (2012). Finite element prediction of cartilage contact stresses in normal human hips. *Journal of Orthopaedic Research, 30*(7), 1133–1139.

Haut, T. L., Hull, M. L., & Howell, S. M. (1997). A high accuracy three-dimensional coordinate digitizing system for reconstructing the geometry of diarthrodial joints. *ASME-PUBLICATIONS-BED, 36,* 17–18.

Haut, R. C., Ide, T. M., & De Camp, C. E. (1995). Mechanical responses of the rabbit patello-femoral joint to blunt impact. *Transactions-American Society of Mechanical Engineers Journal of Biomechanical Engineering, 117,* 402–408.

Hayes, W. C., & Bouxsein, M. L. (1991). Biomechanics of cortical and trabecular bone: implications for assessment of fracture risk. *Basic Orthopaedic Biomechanics, 2,* 69–111.

Hayes, W. C., & Mockros, L. F. (1971). Viscoelastic properties of human articular cartilage. *Journal of Applied Physiology, 31*(4), 562–568.

Hayes, W. C., et al. (1972). A mathematical analysis for indentation tests of articular cartilage. *Journal of Biomechanics, 5*(5), 541–551.

Hefzy, M. S., Grood, E. S., & Zoghi, M. (1987). An axisymmetric finite element model of the meniscus. In *1987 Advances in Bioengineering* (pp. 51–52).

Herzog, W. (2004). Effect of fluid boundary conditions on joint contact mechanics and applications to the modeling of osteoarthritic joints. *Journal of Biomechanical Engineering, 126,* 220.

Herzog, W., et al. (1998). Material and functional properties of articular cartilage and patellofemoral contact mechanics in an experimental model of osteoarthritis. *Journal of Biomechanics, 31*(12), 1137–1145.

Hirokawa, S., & Tsuruno, R. (2000). Three-dimensional deformation and stress distribution in an analytical/computational model of the anterior cruciate ligament. *Journal of Biomechanics, 33*(9), 1069–1077.

Hodge, W. A., et al. (1986). Contact pressures in the human hip joint measured in vivo. *Proceedings of the National Academy of Sciences, 83*(9), 2879–2883.

Holmes, M. H. (1986). Finite deformation of soft tissue: analysis of a mixture model in uni-axial compression. *Journal of Biomechanical Engineering, 108*(4), 372–381.

Holzapfel, G. A. (2002). Nonlinear solid mechanics: A continuum approach for engineering science. *Meccanica, 37*(4), 489–490.

Hsieh, H.-H., & Walker, P. S. (1976). Stabilizing mechanisms of the loaded and unloaded knee joint. *Journal of Bone and Joint Surgery. American Volume, 58*(1), 87–93.

Izaham, R. M. A. R., et al. (2012). Finite element analysis of Puddu and Tomofix plate fixation for open wedge high tibial osteotomy. *Injury, 43*(6), 898–902.

Johansson, T., Meier, P., & Blickhan, R. (2000). A finite-element model for the mechanical analysis of skeletal muscles. *Journal of Theoretical Biology, 206*(1), 131–149.

Jolivet, E., Poméro, V., & Skalli, W. (2001). Finite element model of muscle. In *International Symposium on Computer Methods in Biomechanics and Biomedical Engineering*.

Julkunen, P., et al. (2007). Characterization of articular cartilage by combining microscopic analysis with a fibril-reinforced finite-element model. *Journal of Biomechanics, 40*(8), 1862–1870.

Kahle, W., et al. (1998). Système nerveux et organes des sens. Flammarion Médecine-Sciences, Paris.

Kanamori, A., et al. (1998). The forces in the ACL and knee kinematics during the clinical "Pivot Shift" test. In *Transactions of the Annual Meeting-Orthopaedic Research Society* (p. 816). Orthopaedic research society.

Katsamanis, F., & Raftopoulos, D. D. (1990). Determination of mechanical properties of human femoral cortical bone by the Hopkinson bar stress technique. *Journal of Biomechanics, 23*(11), 1173–1184.

Keer, L. M., Lewis, J. L., & Vithoontien, V. (1990). An analytical model of joint contact. *Journal of Biomechanical Engineering, 112,* 407.

Kempson, G. E. (1979). Mechanical properties of articular cartilage. In M. A. R. Freeman (Ed.), *Adult articular cartilage* (pp. 313–414). Kent: Pitman medical.

Kempson, G. E. (1980). The mechanical properties of articular cartilage. *The Joints and Synovial Fluid, 2,* 177–238.

Kempson, G. E., et al. (1976). The effects of proteolytic enzymes on the mechanical properties of adult human articular cartilage. *Biochimica et Biophysica Acta (BBA)-General Subjects, 428* (3), 741–760.

Kiapour, A. M., et al. (2014). The effect of ligament modeling technique on knee joint kinematics: A finite element study. *Applied Mathematics, 4*(5A), 91.

Kluess, D., et al. (2009). A convenient approach for finite-element-analyses of orthopaedic implants in bone contact: modeling and experimental validation. *Computer Methods and Programs in Biomedicine, 95*(1), 23–30.

Kohn, D., & Moreno, B. (1995). Meniscus insertion anatomy as a basis for meniscus replacement: a morphological cadaveric study. *Arthroscopy: The Journal of Arthroscopic & Related Surgery, 11*(1), 96–103.

Korhonen, R. K., et al. (2003). Fibril reinforced poroelastic model predicts specifically mechanical behavior of normal, proteoglycan depleted and collagen degraded articular cartilage. *Journal of Biomechanics, 36*(9), 1373–1379.

Kurosawa, H., et al. (1985). Geometry and motion of the knee for implant and orthotic design. *Journal of Biomechanics, 18*(7), 487493–491499.

Kusayama, T., et al. (1994). Anatomical and biomechanical characteristics of human meniscofemoral ligaments. *Knee Surgery, Sports Traumatology, Arthroscopy, 2*(4), 234–237.

Kwak, S. D., Blankevoort, L., & Ateshian, G. A. (2000). A mathematical formulation for 3D quasi-static multibody models of diarthrodial joints. *Computer Methods in Biomechanics and Biomedical Engineering, 3*(1), 41–64.

Laasanen, M. S., et al. (2003). Biomechanical properties of knee articular cartilage. *Biorheology, 40*(1, 2, 3), 133–140.

Lechner, K., Hull, M. L., & Howell, S. M. (2000). Is the circumferential tensile modulus within a human medial meniscus affected by the test sample location and cross-sectional area? *Journal of Orthopaedic Research, 18*(6), 945–951.

Lengsfeld, M., et al. (1998). Comparison of geometry-based and CT voxel-based finite element modelling and experimental validation. *Medical Engineering & Physics, 20*(7), 515–522.

LeRoux, M. A., & Setton, L. A. (2002). Experimental and biphasic FEM determinations of the material properties and hydraulic permeability of the meniscus in tension. *Journal of Biomechanical Engineering, 124*(3), 315–321.

Li, L. P., Buschmann, M. D., & Shirazi-Adl, A. (2000). A fibril reinforced nonhomogeneous poroelastic model for articular cartilage: Inhomogeneous response in unconfined compression. *Journal of Biomechanics, 33*(12), 1533–1541.

Li, L., Cheung, J. T. M., & Herzog, W. (2009). Three-dimensional fibril-reinforced finite element model of articular cartilage. *Medical & Biological Engineering & Computing, 47*(6), 607.

Li, L. P., & Herzog, W. (2004). The role of viscoelasticity of collagen fibers in articular cartilage: Theory and numerical formulation. *Biorheology, 41*(3–4), 181–194.

Li, G., Lopez, O., & Rubash, H. (2001a). Variability of a three-dimensional finite element model constructed using magnetic resonance images of a knee for joint contact stress analysis. *Journal of Biomechanical Engineering, 123*(4), 341–346.

Li, G., Orlando, L., & Harry, H. (2001b). Variability of a three dimensional finite element model constructed using magnetic resonance images of a knee for joint contact stress analysis. *Journal of Biomechanical Engineering-Transactions of the Asme, 123*(4), 341–346.

Li, G., Suggs, J., & Gill, T. (2002). The effect of anterior cruciate ligament injury on knee joint function under a simulated muscle load: A three-dimensional computational simulation. *Annals of Biomedical Engineering, 30*(5), 713–720.

Li, G., et al. (1998). Effect of combined axial compressive and anterior tibial loads on in situ forces in the anterior cruciate ligament: A porcine study. *Journal of Orthopaedic Research, 16*(1), 122–127.

Li, G., et al. (1999a). A validated three-dimensional computational model of a human knee joint. *Journal of Biomechanical Engineering, 121*(6), 657–662.

Li, L. P., et al. (1999b). Nonlinear analysis of cartilage in unconfined ramp compression using a fibril reinforced poroelastic model. *Clinical Biomechanics, 14*(9), 673–682.

Li, G., et al. (2005a). In vivo articular cartilage contact kinematics of the knee an investigation using dual-orthogonal fluoroscopy and magnetic resonance image-based computer models. *The American Journal of Sports Medicine, 33*(1), 102–107.

Li, L. P., et al. (2005b). The role of viscoelasticity of collagen fibers in articular cartilage: Axial tension versus compression. *Medical Engineering & Physics, 27*(1), 51–57.

Limbert, G., Middleton, J., & Taylor, M. (2004). Finite element analysis of the human ACL subjected to passive anterior tibial loads. *Computer Methods in Biomechanics and Biomedical Engineering, 7*(1), 1–8.

Little, R. B., et al. (1986). A three-dimensional finite element analysis of the upper Tibia. *Journal of Biomechanical Engineering, 108*(2), 111–119.

Łuczkiewicz, P., et al. (2016). The influence of articular cartilage thickness reduction on meniscus biomechanics. *PLoS ONE, 11*(12), e0167733.

Luo, Z.-P., et al. (1998). Mechanical environment associated with rotator cuff tears. *Journal of Shoulder and Elbow Surgery, 7*(6), 616–620.

Makris, E. A., Hadidi, P., & Athanasiou, K. A. (2011). The knee meniscus: structure–function, pathophysiology, current repair techniques, and prospects for regeneration. *Biomaterials, 32*(30), 7411–7431.

Markolf, K. L., Mensch, J. S., & Amstutz, H. C. (1976). Stiffness and laxity of the knee–the contributions of the supporting structures. A quantitative in vitro study. *Journal of Bone and Joint Surgery. American Volume, 58*(5), 583–594.

Markolf, K. L., Wascher, D. C., & Finerman, G. A. (1993). Direct in vitro measurement of forces in the cruciate ligaments. Part II: The effect of section of the posterolateral structures. *Journal of Bone and Joint Surgery. American Volume, 75*(3), 387–394.

Markolf, K. L., et al. (1981). The role of joint load in knee stability. *Journal of Bone and Joint Surgery, 63*(4), 570–585.

Markolf, K. L., et al. (1995). Combined knee loading states that generate high anterior cruciate ligament forces. *Journal of Orthopaedic Research, 13*(6), 930–935.

Martin, R. B., Burr, D. B., & Sharkey, N. A. (1998). *Skeletal tissue mechanics.* Springer.

Mavčič, B., et al. (2000). Weight bearing area during gait in normal and dysplastic hips. *Pflügers Archiv-European Journal of Physiology, 439*(7), R213–R214.

McDermott, I. D., Masouros, S. D., & Amis, A. A. (2008). Biomechanics of the menisci of the knee. *Current Orthopaedics, 22*(3), 193–201.

McDermott, I. D., et al. (2004). An anatomical study of meniscal allograft sizing. *Knee Surgery, Sports Traumatology, Arthroscopy, 12*(2), 130–135.

Meakin, J. R., et al. (2003). Finite element analysis of the meniscus: the influence of geometry and material properties on its behaviour. *The Knee, 10*(1), 33–41.

Mesfar, W., & Shirazi-Adl, A. (2005). Biomechanics of the knee joint in flexion under various quadriceps forces. *The Knee, 12*(6), 424–434.

Mina, C., et al. (2008). High tibial osteotomy for unloading osteochondral defects in the medial compartment of the knee. *The American Journal of Sports Medicine, 36*(5), 949–955.

Miyoshi, S., et al. (2002a). Analysis of the shape of the tibial tray in total knee arthroplasty using a three dimension finite element model. *Clinical Biomechanics, 17*(7), 521–525.

Miyoshi, S., et al. (2002b). Analysis of the shape of the tibial tray in total knee arthroplasty using a three dimension finite element model. *Clinical biomechanics (Bristol, Avon), 17*(7), 521–525. Available at: http://www.ncbi.nlm.nih.gov/pubmed/12206943.

Moglo, K. E., & Shirazi-Adl, A. (2003). Biomechanics of passive knee joint in drawer: Load transmission in intact and ACL-deficient joints. *The Knee, 10*(3), 265–276.

Mommersteeg, T. J. A., et al. (1996). A global verification study of a quasi-static knee model with multi-bundle ligaments. *Journal of Biomechanics, 29*(12), 1659–1664.

Mononen, M. E., et al. (2012). Effect of superficial collagen patterns and fibrillation of femoral articular cartilage on knee joint mechanics-a 3D finite element analysis. *Journal of Biomechanics, 45*(3), 579–587.

Mooney, M. (1940). A theory of large elastic deformation. *Journal of Applied Physics, 11*(9), 582–592.

Moore, S. M., et al. (2010). The glenohumeral capsule should be evaluated as a sheet of fibrous tissue: a validated finite element model. *Annals of Biomedical Engineering, 38*(1), 66–76.

Mootanah, R., et al. (2014). Development and validation of a computational model of the knee joint for the evaluation of surgical treatments for osteoarthritis. *Computer Methods in Biomechanics and Biomedical Engineering, 17*(13), 1502–1517.

Morimoto, Y., et al. (2009). Tibiofemoral joint contact area and pressure after single-and double-bundle anterior cruciate ligament reconstruction. *Arthroscopy: The Journal of Arthroscopic & Related Surgery, 25*(1), 62–69.

Mow, V. C., & Guo, X. E. (2002). Mechano-electrochemical properties of articular cartilage: their inhomogeneities and anisotropies. *Annual Review of Biomedical Engineering, 4*(1), 175–209.

Mow, V. C., Lai, W. M., & Holmes, M. H. (1982). Advanced theoretical and experimental techniques in cartilage research. In *Biomechanics: Principles and applications* (pp. 47–74). Springer.

Nakajima, T., et al. (1994). Histologic and biomechanical characteristics of the supraspinatus tendon: Reference to rotator cuff tearing. *Journal of Shoulder and Elbow Surgery, 3*(2), 79–87.

Namani, R., Simha, N. K., & Lewis, J. L. (2003). Nonlinear elastic parameters of articular cartilage. In *Summer Bioengineering Conference, June* (pp. 25–29).

Netravali, N. A., et al. (2011). The effect of kinematic and kinetic changes on meniscal strains during gait. *Journal of Biomechanical Engineering, 133*(1), 11006.

Newberry, W. N., Zukosky, D. K., & Haut, R. C. (1997). Subfracture insult to a knee joint causes alterations in the bone and in the functional stiffness of overlying cartilage. *Journal of Orthopaedic Research, 15*(3), 450–455.

O'Connor, J. J. (1993). Can muscle co-contraction protect knee ligaments after injury or repair? *Bone & Joint Journal, 75*(1), 41–48.

Oloyede, A., Flachsmann, R., & Broom, N. D. (1992). The dramatic influence of loading velocity on the compressive response of articular cartilage. *Connective Tissue Research, 27*(4), 211–224.

Oonishi, H., Isha, H., & Hasegawa, T. (1983). Mechanical analysis of the human pelvis and its application to the artificial hip joint—by means of the three dimensional finite element method. *Journal of Biomechanics, 16*(6), 427–444.

Pandy, M. G., & Sasaki, K. (1998). A three-dimensional musculoskeletal model of the human knee joint. Part 2: analysis of ligament function. *Computer Methods in Biomechanics and Bio Medical Engineering, 1*(4), 265–283.

Pandy, M. G., Sasaki, K., & Kim, S. (1997). A three-dimensional musculoskeletal model of the human knee joint. Part 1: theoretical construction. *Computer Methods in Biomechanics and Bio Medical Engineering, 1*(2), 87–108.

Peña, E., Calvo, B., et al. (2005a). Finite element analysis of the effect of meniscal tears and meniscectomies on human knee biomechanics. *Clinical Biomechanics, 20*(5), 498–507.

Peña, E., Martinez, M. A., et al. (2005b). A finite element simulation of the effect of graft stiffness and graft tensioning in ACL reconstruction. *Clinical Biomechanics, 20*(6), 636–644.

Peña, E., et al. (2006a). A three-dimensional finite element analysis of the combined behavior of ligaments and menisci in the healthy human knee joint. *Journal of Biomechanics, 39*(9), 1686–1701.

Peña, E., et al. (2006b). Why lateral meniscectomy is more dangerous than medial meniscectomy. A finite element study. *Journal of Orthopaedic Research, 24*(5), 1001–1010.

Peña, E., et al. (2007). Effect of the size and location of osteochondral defects in degenerative arthritis. A finite element simulation. *Computers in Biology and Medicine, 37*(3), 376–387.

Peña, E., et al. (2008). Computer simulation of damage on distal femoral articular cartilage after meniscectomies. *Computers in Biology and Medicine, 38*(1), 69–81.

Penrose, J. M. T., et al. (2002). Development of an accurate three-dimensional finite element knee model. *Computer Methods in Biomechanics & Biomedical Engineering, 5*(4), 291–300.

Perie, D., & Hobatho, M. C. (1998). In vivo determination of contact areas and pressure of the femorotibial joint using non-linear finite element analysis. *Clinical Biomechanics, 13*(6), 394–402.

Piazza, S. J., & Delp, S. L. (2001). Three-dimensional dynamic simulation of total knee replacement motion during a step-up task. *Journal of Biomechanical Engineering, 123*(6), 599–606.

Pioletti, D. P., & Rakotomanana, L. R. (2000). Non-linear viscoelastic laws for soft biological tissues. *European Journal of Mechanics A-Solids, 19* (LBO-ARTICLE-2000-002), 749–759.

Pioletti, D. P., et al. (1998). Viscoelastic constitutive law in large deformations: application to human knee ligaments and tendons. *Journal of Biomechanics, 31*(8), 753–757.

Ramaniraka, N. A., Saunier, P., & Siegrist, O. (2005a). Effects of intra-articular and extra-articular procedures in anterior cruciate ligament (ACL) reconstruction. *Computer Methods in Biomechanics and Biomedical Engineering, 8*(S1), 231–232.

Ramaniraka, N. A., Terrier, A., et al. (2005b). Effects of the posterior cruciate ligament reconstruction on the biomechanics of the knee joint: A finite element analysis. *Clinical Biomechanics, 20*(4), 434–442.

Ramaniraka, N. A., et al. (2007). Biomechanical evaluation of intra-articular and extra-articular procedures in anterior cruciate ligament reconstruction: A finite element analysis. *Clinical Biomechanics, 22*(3), 336–343.

Rapperport, D. J., Carter, D. R., & Schurman, D. J. (1985). Contact finite element stress analysis of the hip joint. *Journal of Orthopaedic Research, 3*(4), 435–446.

Reilly, D. T., Burstein, A. H., & Frankel, V. H. (1974). The elastic modulus for bone. *Journal of Biomechanics, 7*(3), 271–275.

Repo, R. U., & Finlay, J. B. (1977). Survival of articular cartilage after controlled impact. *Journal of Bone and Joint Surgery. American Volume, 59*(8), 1068–1076.

Reuben, J. D., et al. (1986). Three-dimensional kinematics of normal and cruciate deficient knees —A dynamic in-vitro experiment. *Transactions of the Orthopedic Research Society, 11,* 385.

Rho, J. Y., Ashman, R. B., & Turner, C. H. (1993). Young's modulus of trabecular and cortical bone material: ultrasonic and microtensile measurements. *Journal of Biomechanics, 26*(2), 111–119.

Rice, J. C., Cowin, S. C., & Bowman, J. A. (1988). On the dependence of the elasticity and strength of cancellous bone on apparent density. *Journal of Biomechanics, 21*(2), 155–168.

Richard, F., Villars, M., & Thibaud, S. (2013). Viscoelastic modeling and quantitative experimental characterization of normal and osteoarthritic human articular cartilage using indentation. *Journal of the Mechanical Behavior of Biomedical Materials, 24,* 41–52.

Rivlin, R. S. (1948). Large elastic deformations of isotropic materials. IV. Further developments of the general theory. *Philosophical Transactions of the Royal Society of London A: Mathematical, Physical and Engineering Sciences, 241*(835), 379–397.

Roth, V. (1977). *Two problems in articular biomechanics: I. Finite element simulation for contact problems of articulations, II. Age dependent tensile properties.* Ph.D. thesis, Rensselaer Polytechnic Institute, Troy, New York.

Rovick, J., et al. (1986). *The influence of the ACL on the motion of the knee.* Sun Valley, Idaho: AOSSM.

Rudy, T. W., et al. (1996). A combined robotic/universal force sensor approach to determine in situ forces of knee ligaments. *Journal of Biomechanics, 29*(10), 1357–1360.

Russell, M. E., et al. (2006). Cartilage contact pressure elevations in dysplastic hips: A chronic overload model. *Journal of Orthopaedic Surgery and Research, 1*(1), 1.

Schreppers, G., Sauren, A., & Huson, A. (1990). A numerical model of the load transmission in the tibio-femoral contact area. *Proceedings of the Institution of Mechanical Engineers. Part H, Journal of Engineering in Medicine, 204*(1), 53–59.

Schütz, U. H. W., et al. (2012). The Transeurope footrace project: Longitudinal data acquisition in a cluster randomized mobile MRI observational cohort study on 44 endurance runners at a 64-stage 4,486 km transcontinental ultramarathon. *BMC Medicine, 10*(1), 78.

Setton, L. A., Zhu, W., & Mow, V. C. (1993). The biphasic poroviscoelastic behavior of articular cartilage: role of the surface zone in governing the compressive behavior. *Journal of Biomechanics, 26*(4), 581–592.

Setton, L. A., et al. (1994). Mechanical properties of canine articular cartilage are significantly altered following transection of the anterior cruciate ligament. *Journal of Orthopaedic Research, 12*(4), 451–463.

Shaffer, B., et al. (2000). Preoperative sizing of meniscal allografts in meniscus transplantation. *The American Journal of Sports Medicine, 28*(4), 524–533.

Shelburne, K. B., & Pandy, M. G. (1997). A musculoskeletal model of the knee for evaluating ligament forces during isometric contractions. *Journal of Biomechanics, 30*(2), 163–176.

Shepherd, D. E., & Seedhom, B. B. (1999). The 'instantaneous' compressive modulus of human articular cartilage in joints of the lower limb. *Rheumatology, 38*(2), 124–132.

Shirazi, R., & Shirazi-Adl, A. (2009a). Analysis of partial meniscectomy and ACL reconstruction in knee joint biomechanics under a combined loading. *Clinical Biomechanics, 24*(9), 755–761.

Shirazi, R., & Shirazi-Adl, A. (2009b). Computational biomechanics of articular cartilage of human knee joint: effect of osteochondral defects. *Journal of Biomechanics, 42*(15), 2458–2465.

Shirazi, R., Shirazi-Adl, A., & Hurtig, M. (2008). Role of cartilage collagen fibrils networks in knee joint biomechanics under compression. *Journal of Biomechanics, 41*(16), 3340–3348.

Shriram, D., et al. (2017). Evaluating the effects of material properties of artificial meniscal implant in the human knee joint using finite element analysis. *Scientific Reports, 7*(1), 6011.

Skaggs, D. L., Warden, W. H., & Mow, V. C. (1994). Radial tie fibers influence the tensile properties of the bovine medial meniscus. *Journal of Orthopaedic Research, 12*(2), 176–185.

Song, Y., et al. (2004). A three-dimensional finite element model of the human anterior cruciate ligament: a computational analysis with experimental validation. *Journal of Biomechanics, 37*(3), 383–390.

Spilker, R. L., Donzelli, P. S., & Mow, V. C. (1992). A transversely isotropic biphasic finite element model of the meniscus. *Journal of Biomechanics, 25*(9), 1027–1045.

Sweigart, M. A., et al. (2004). Intraspecies and interspecies comparison of the compressive properties of the medial meniscus. *Annals of Biomedical Engineering, 32*(11), 1569–1579.

Terrier, A., et al. (2007). Effect of supraspinatus deficiency on humerus translation and glenohumeral contact force during abduction. *Clinical Biomechanics, 22*(6), 645–651.

Thompson, R. C., et al. (1991). Osteoarthrotic changes after acute transarticular load. An animal model. *Journal of Bone and Joint Surgery. American Volume, 73*(7), 990–1001.

Tissakht, M., & Ahmed, A. M. (1995). Tensile stress-strain characteristics of the human meniscal material. *Journal of Biomechanics, 28*(4), 411–422.

Trad, Z., Barkaoui, A., & Chafra, M. (2017). A three dimensional finite element analysis of mechanical stresses in the human knee joint: Problem of cartilage destruction. In *Journal of Biomimetics, Biomaterials and Biomedical Engineering* (pp. 29–39). Trans Tech Publications.

Trent, P. S., Walker, P. S., & Wolf, B. (1976). Ligament length patterns, strength, and rotational axes of the knee joint. *Clinical Orthopaedics and Related Research, 117*, 263–270.

Turner, S. T., & Engin, A. E. (1993). Three-body segment dynamic model of the human knee. *Journal Biomechanical Engineering, 115*(4), 350–356.

Un, K. (2001). An evaluation of three-dimensional diarthrodial joint contact using penetration data and the finite element method. *Journal Biomechanical Engineering, 123*(4), 333-340 (Mar 26, 2001) (8 pages) doi:https://doi.org/10.1115/1.1384876

Un, K., & Spilker, R. L. (2006). A penetration-based finite element method for hyperelastic 3D biphasic tissues in contact: Part 1–Derivation of contact boundary conditions. *Journal of Biomechanical Engineering, 128*(1), 124–130.

Vadher, S. P., et al. (2006). Finite element modeling following partial meniscectomy: Effect of various size of resection. In *Engineering in Medicine and Biology Society, 2006. EMBS'06. 28th Annual International Conference of the IEEE* (pp. 2098–2101). IEEE.

Vaziri, A., et al. (2008). Influence of meniscectomy and meniscus replacement on the stress distribution in human knee joint. *Annals of Biomedical Engineering, 36*(8), 1335–1344.

Vener, M. J., et al. (1992). Subchondral damage after acute transarticular loading: An in vitro model of joint injury. *Journal of Orthopaedic Research, 10*(6), 759–765.

Viswanath, B., et al. (2007). Mechanical properties and anisotropy in hydroxyapatite single crystals. *Scripta Materialia, 57*(4), 361–364.

von Eisenhart-Rothe, R., et al. (1997). Direct comparison of contact areas, contact stress and subchondral mineralization in human hip joint specimens. *Anatomy and Embryology, 195*(3), 279–288.

Walia, P., et al. (2013). Theoretical model of the effect of combined glenohumeral bone defects on anterior shoulder instability: A finite element approach. *Journal of Orthopaedic Research, 31* (4), 601–607.

Walker, P. S., & Hajek, J. V. (1972). The load-bearing area in the knee joint. *Journal of Biomechanics, 5*(6), 581IN3585–584IN5589.

Wan, C., Hao, Z., & Wen, S. (2013). The effect of the variation in ACL constitutive model on joint kinematics and biomechanics under different loads: A finite element study. *Journal of Biomechanical Engineering, 135*(4), 41002.

Wang, Y., Fan, Y., & Zhang, M. (2014). Comparison of stress on knee cartilage during kneeling and standing using finite element models. *Medical Engineering & Physics, 36*(4), 439–447.

Wang, C., & Walker, P. S. (1974). Rotatory laxity of the human knee joint. *Journal of Bone and Joint Surgery. American Volume, 56*(1), 161–170.

Wang, C.-J., Walker, P. S., & Wolf, B. (1973). The effects of flexion and rotation on the length patterns of the ligaments of the knee. *Journal of Biomechanics, 6*(6), 587IN1593–592IN4596.

Wascher, D. C., et al. (1993). Direct in vitro measurement of forces in the cruciate ligaments. Part I: The effect of multiplane loading in the intact knee. *Journal of Bone and Joint Surgery. American Volume, 75*(3), 377–386.

Wei, H.-W., et al. (2005). The influence of mechanical properties of subchondral plate, femoral head and neck on dynamic stress distribution of the articular cartilage. *Medical Engineering & Physics, 27*(4), 295–304.

Weiss, J. A., & Gardiner, J. C. (2001). Computational modeling of ligament mechanics. *Critical Reviews^{TM} in Biomedical Engineering, 29*(3).

Weiss, J. A., Gardiner, J. C., & Bonifasi-Lista, C. (2002). Ligament material behavior is nonlinear, viscoelastic and rate-independent under shear loading. *Journal of Biomechanics, 35*(7), 943–950.

Weiss, J. A., Maker, B. N., & Govindjee, S. (1996). Finite element implementation of incompressible, transversely isotropic hyperelasticity. *Computer Methods in Applied Mechanics and Engineering, 135*(1–2), 107–128.

Weiss, J. A., Maker, B. N., & Schauer, D. A. (1995). Treatment of initial stress in hyperelastic finite element models of soft tissues. *ASME-PUBLICATIONS-BED, 29,* 105.

Westermann, R. W., Wolf, B. R., & Elkins, J. M. (2013). Effect of ACL reconstruction graft size on simulated Lachman testing: A finite element analysis. *The Iowa Orthopaedic Journal, 33,* 70.

Whipple, R., Wirth, C. R., & Mow, V. C. (1984). Mechanical properties of the meniscus. In *1984 Advances in Bioengineering* (pp. 32–33).

Wilson, W., et al. (2003). Pathways of load-induced cartilage damage causing cartilage degeneration in the knee after meniscectomy. *Journal of Biomechanics, 36*(6), 845–851.

Wilson, W., et al. (2004). Stresses in the local collagen network of articular cartilage: a poroviscoelastic fibril-reinforced finite element study. *Journal of Biomechanics, 37*(3), 357–366.

Wilson, W., et al. (2005). The role of computational models in the search for the mechanical behavior and damage mechanisms of articular cartilage. *Medical Engineering & Physics, 27* (10), 810–826.

Wismans, J. A. C., et al. (1980). A three-dimensional mathematical model of the knee-joint. *Journal of Biomechanics, 13*(8), 677681–679685.

Woo, S. L. Y., et al. (1998). Biomechanics of the ACL: Measurements of in situ force in the ACL and knee kinematics. *The Knee, 5*(4), 267–288.

Woo, S. L. Y., et al. (1999). Biomechanics of knee ligaments. *The American Journal of Sports Medicine, 27*(4), 533–543.

Wu, J. Z., & Herzog, W. (2000). Finite element simulation of location-and time-dependent mechanical behavior of chondrocytes in unconfined compression tests. *Annals of Biomedical Engineering, 28*(3), 318–330.

Wu, J. Z., Herzog, W., & Epstein, M. (1997). Evaluation of the finite element software ABAQUS for biomechanical modelling of biphasic tissues. *Journal of Biomechanics, 31*(2), 165–169.

Yagi, M., et al. (2002). Biomechanical analysis of an anatomic anterior cruciate ligament reconstruction. *The American journal of sports medicine, 30*(5), 660–666.

Yamamoto, K., Hirokawa, S., & Kawada, T. (1998). Strain distribution in the ligament using photoelasticity. A direct application to the human ACL. *Medical Engineering & Physics, 20*(3), 161–168.

Yang, K. H., & Radin, E. L. (1990). A dynamic finite element analysis of impulsive loading of the extension-splinted rabbit knee. *Journal of Biomechanical Engineering, 112,* 119.

Yang, N. H., et al. (2010a). Effect of frontal plane tibiofemoral angle on the stress and strain at the knee cartilage during the stance phase of gait. *Journal of Orthopaedic Research, 28*(12), 1539–1547.

Yang, N. H., et al. (2010b). Protocol for constructing subject-specific biomechanical models of knee joint. *Computer Methods in Biomechanics and Biomedical Engineering, 13*(5), 589–603.

Yao, J., Funkenbusch, P. D., et al. (2006a). Sensitivities of medial meniscal motion and deformation to material properties of articular cartilage, meniscus and meniscal attachments using design of experiments methods. *Journal of Biomechanical Engineering, 128*(3), 399–408.

Yao, J., Snibbe, J., et al. (2006b). Stresses and strains in the medial meniscus of an ACL deficient knee under anterior loading: A finite element analysis with image-based experimental validation. *Journal of Biomechanical Engineering, 128*(1), 135–141.

Yucesoy, C. A., et al. (2002). Three-dimensional finite element modeling of skeletal muscle using a two-domain approach: linked fiber-matrix mesh model. *Journal of Biomechanics, 35*(9), 1253–1262.

Zahnert, T., et al. (2000). Experimental investigations of the use of cartilage in tympanic membrane reconstruction. *Otology & Neurotology, 21*(3), 322–328.

Zaki, M., Saad, F., & Al-Ebiary, M. N. (2002). Influence of Charnley hip neck-angle inclination on the stresses at stem/cement and bone/cement interfaces. *Bio-Medical Materials and Engineering, 12*(4), 411–421.

Zavatsky, A. B., & O'Connor, J. J. (1993). Ligament forces at the knee during isometric quadriceps contractions. *Proceedings of the Institution of Mechanical Engineers. Part H, Journal of Engineering in Medicine, 207*(1), 7–18.

Zhang, H., et al. (1999a). Magnetic resonance image based 3D poroelastic finite element model of tibio-menisco-femoral contact. In *23rd Proceedings of the American Society of Biomechanics* (pp. 198–199).

Zhang, H., et al. (1999b). Damage to rabbit femoral articular cartilage following direct impacts of uniform stresses: An in vitro study. *Clinical Biomechanics, 14*(8), 543–548.

Zheng, K. (2014). The effect of high tibial osteotomy correction angle on cartilage and meniscus loading using finite element analysis. Ph.D. thesis, School of Aerospace, Mechanical and Mechatronic Engineering, the University of Sydney, Australia.

Zheng, K. K., et al. (2014). Magnetic resonance imaging (MRI) based finite element modeling for analyzing the influence of material properties on menisci responses. In *Applied Mechanics and Materials* (pp. 305–309). Trans Tech Publications.

Zhu, G.-D., et al. (2015). Finite element analysis of mobile-bearing unicompartmental knee arthroplasty: The influence of tibial component coronal alignment. *Chinese Medical Journal, 128*(21), 2873.

Zielinska, B., & Donahue, T. L. H. (2006). 3D finite element model of meniscectomy: Changes in joint contact behavior. *Journal of Biomechanical Engineering, 128*(1), 115–123.

Zuppinger, H. (1904). Die aktive Flexion im unbelasteten Kniegelenk. *Anatomische Hefte, 25*(3), 701–764.

Zysset, P. K., et al. (1999). Elastic modulus and hardness of cortical and trabecular bone lamellae measured by nanoindentation in the human femur. *Journal of Biomechanics, 32*(10), 1005–1012.

Chapter 2
Finite Element Analysis Applications in Biomechanical Studies of the Knee Joint

The development of sophisticated 3D FE models through MRI techniques enables us to precisely capture the patient-specific geometries of both hard and soft tissues in the region of interest (RoI), in order to more precisely simulate complicated tissue responses, thereby reflecting more realistic biomechanical behaviors. In the past decades, extensive studies have developed FE models and have coupled the FE model with in vivo kinematic data to analyse true tissue deformation (Halloran et al. 2010). This has resulted in a more convincing simulation and prediction of the loading condition in FEA.

The following section reviews studies published in the literature that have used FE techniques to analyse the stress and strain distribution at the knee joint under different loading conditions so as to investigate the effects of ligament injury, meniscus injury, cartilage disease, HTO and normal knee joint contact mechanism.

Several FEA studies have demonstrated the validity and effectiveness of knee biomechanics under simulated realistic loading under either physiological or non-physiological conditions. Therefore, this review is only focused on those previous studies that developed 3D FE models with attention on analyzing the biomechanics of the knee joint.

2.1 Current FEA Applications on Ligament Injury

By analyzing many previous FEA studies, validation of the FE model of the knee joint was considered a primary step for its critical role in understanding the biomechanical details. As mentioned in the previous section, Li et al. (1999) created the first 3D FE model of the human knee joint to be validated using experimental data. The geometry of the joint model was obtained from an MRI of a cadaveric knee specimen. In the finite element model (FEM) built, cartilage was modeled as an elastic material, ligaments were represented as nonlinear elastic springs, and menisci were simulated by equivalent-resistance springs. The reference lengths of

© The Author(s), under exclusive licence to Springer International Publishing AG, part of Springer Nature 2018
Z. Trad et al., *FEM Analysis of the Human Knee Joint*, SpringerBriefs in Applied Sciences and Technology, https://doi.org/10.1007/978-3-319-74158-1_2

the ligaments and stiffness of the meniscus springs were estimated using an optimization procedure that involved the minimization of the differences between the kinematics predicted by the model and those obtained experimentally.

The purpose of this study was to predict the ligament forces and kinematics of the knee in response to internal-external moments, and the computed results were compared with published experimental data as a validation of the model. Furthermore, as a sample application, the validated model was used to investigate the effect of removing the menisci on the rotational laxity of the knee joint. As a result, the methodology and protocol of FE modeling presented in the study by Li et al. can be a valuable tool for further analysis of knee joint function and could serve as a step toward the development of more advanced computational knee models.

Hence, the development of a trustworthy model of a complex structure like the knee joint is a greatly challenging process. The difficulties increase even more when a good description of the ligaments' behavior is required. The current FE models of the knee differ in various parameters, such as the study variables, the degree of complexity, the definitions of material models and the loading cases.

A dynamic (Beillas et al. 2004; Godest et al. 2002; Halloran et al. 2005; Baldwin et al. 2012; Penrose et al. 2002; Kiapour et al. 2014), static or quasi-static (Adouni et al. 2012; Dhaher et al. 2010; Donahue et al. 2002; Ellis et al. 2006; Li et al. 1999, 2002; Moglo and Shirazi-Adl 2003; Park et al. 2010; Peña et al. 2006a; Ramaniraka et al. 2005, 2007; Song et al. 2004; Wan et al. 2013; Xie et al. 2009; Yao et al. 2006; Bendjaballah et al. 1997) analysis depends on the choice of the parameters to study.

A literature review article by Gardiner and Weiss (2003) used subject-specific experimental and 3D FE techniques to study the strain distribution in the Medial Collateral Ligament (MCL) under valgus loading during passive flexion and valgus rotation of the knee. Results revealed that the strain distribution within the MCL was non-uniform and changed with the flexion angle. The highest MCL strains occurred at full extension in the posterior region of the MCL proximal to the joint line during valgus loading, which suggests that this region may be most vulnerable to injury under these loading conditions. Thus, this work demonstrates that the developed FE model can predict the complex and non-uniform strain fields that occur in ligaments due to external loading of the joint.

On the other hand, other researchers (Limbert et al. 2004; Pioletti et al. 1998; Song et al. 2004; Hirokawa and Tsuruno 2000) developed a 3D FEM of the Anterior Cruciate Ligament (ACL) in order to study the mechanical behavior of the ACL.

For example, in the study by Song et al. (2004), a validated 3D FEM of the human ACL was developed to calculate the force and stress distribution of the ACL under an anterior tibial load with the knee at full extension. As a result, the anteromedial (AM) and posterolateral (PL) bundles of the ACL shared the force, and the stress distribution was non-uniform within both bundles, with the highest stress localized near the femoral insertion site. The contact and friction caused by the ACL wrapping around the bone during knee motion played the role of transferring the force from the ACL to the bone, and directly affected the force and stress distribution of the ACL.

Later, Fernandes (2014) elaborated a biomechanical analysis of knee behavior after an ACL rupture in order to provide clinically relevant information regarding the force, stress and displacement changes that occur in an ACL-deficient knee. The knee joint model used in this study was obtained from a freely accessible and customizable geometry of the knee joint from the Open Knee project (Erdemir 2013, 2016). All of the knee ligaments were modeled with two hyperelastic constitutive models: the isotropic Marlow (2003) and the anisotropic Holzapfel-Gasser-Ogden (HGO) (Gasser et al. 2006) models. A posterior load of 134 N was applied to the healthy knee, at the Reference Point (RP) that controlled the rigid body motion of the femur at 0°, 15° and 30° of flexion. During application of the force, the flexion-extension degree of freedom (DOF) of the femur was fixed and the tibia was fixed in all DOFs. Thereby, the posterior displacement of the femur relative to the tibia was equivalent to the anterior tibial translation relative to the femur that happens during clinical evaluations of the ACL function.

The results of this article confirmed that the ACL provided the major constraint to anterior tibial motions and enabled restraint of the internal rotation of the tibia. The calculated force sustained by the four ligaments considered also supported these conclusions. Figure 2.1 shows the maximum principal stresses of the soft tissues. When the ACL is absent, the LCL and MCL are the first structures to be required to endure anterior/posterior forces, with obvious implications for their stress state. By comparing the kinematic outcomes of the FE simulations with two material models, it was concluded that the HGO model presented the more accurate findings.

To conclude, this study provides a valuable method for investigating the role of the ACL in the knee. However, some limitations can be indicated. For instance, the

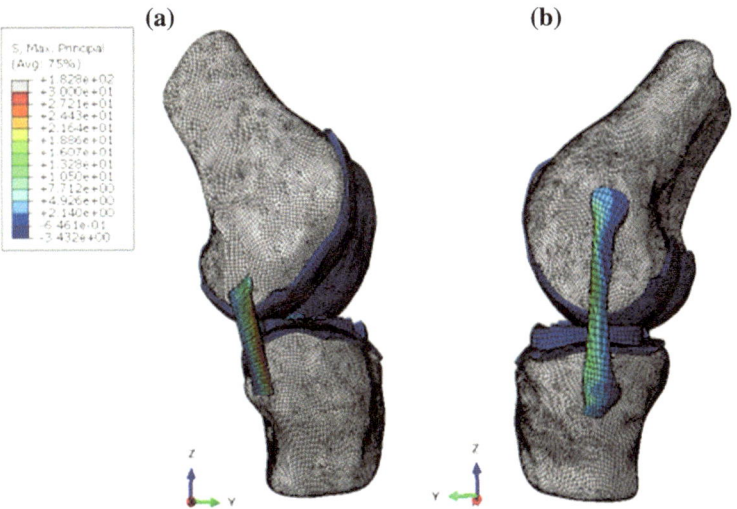

Fig. 2.1 ACL-deficient knee joint at 30° of flexion (**a** lateral view, **b** medial view) (Fernandes 2014)

residual stresses presented on the knee ligaments were not modeled, since the prescription of pre-stresses in the anisotropic hyperelastic model is not allowed. Another shortcoming was that only a posterior femoral load was applied to test the ACL function. The influence other types of movement should also be checked.

2.2 Current FEA Applications on Meniscus Injury

Peña et al. (2005, 2006a, b, 2007, 2008) have carried out a series of FEA studies to investigate the stress and strain behaviors in the soft tissues of a healthy human knee joint, after meniscectomy and with an osteochondral defect. The same 3D FE model of the knee joint based on the MRI was used for all of these studies.

For a study on overall stress distribution in all soft tissues of the knee joint (Peña et al. 2006a), the principal goal was to analyze the combined role of menisci and ligaments in load transmission and stability of the human knee. Therefore, three different load cases were considered. The configuration at full extension served as the reference position. A compression load of 1150 N was applied for the three cases that corresponds to the maximal force in the gait cycle obtained by Sathasivam and Walker (1997) at full extension. A load of 134 N was applied in the anterior-posterior direction for the first case. The two other cases corresponded to rotatory examples, with a valgus torque of 10 Nm in the second load case, adding an anterior load of 134 N in the last one. The results reproduced complex, non-uniform stress and strain distribution that occurs in ligaments (Fig. 2.2), menisci and articular cartilages (Fig. 2.3), and the kinematics of the human knee joint under a physiological external load.

In the studies regarding meniscal tears and meniscectomies (Peña et al. 2005, 2006, 2008), the causes of joint instability with a progressive degenerative arthrosis pathology in articular cartilage have been well investigated. First, the minimal principal stresses corresponding to a compressive load without flexion were

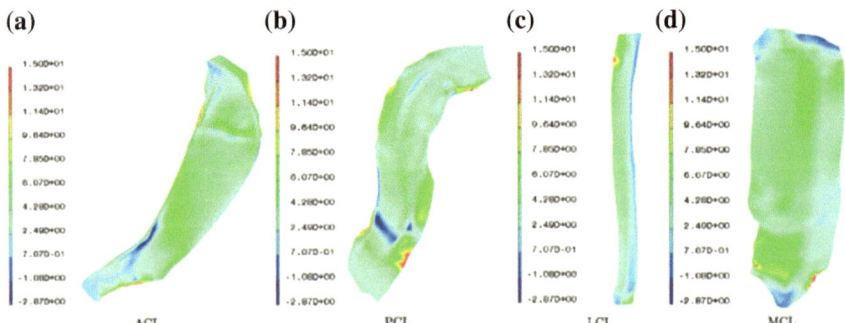

Fig. 2.2 Maximal principal stresses (MPa) in the ligaments in response to a compressive load of 1150 N and an anterior tibial load of 134 N (Peña et al. 2006)

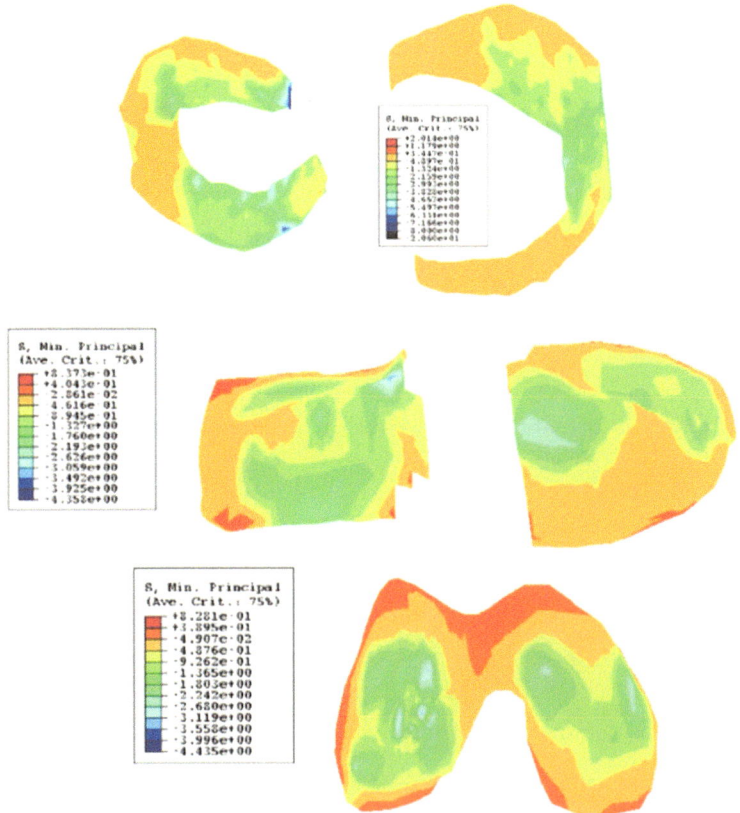

Fig. 2.3 Maximal compressive stresses (MPa) in menisci and articular cartilages in response to a compressive load of 1150 N and an anterior tibial load of 134 N (Peña et al. 2006)

obtained for the posterior zone of the medial meniscus and the corresponding region of the articular cartilage (Fig. 2.4). As a result, under an axial femoral compressive load, the maximal contact stress in the articular cartilage after meniscectomy was about twice that of a healthy joint (Fig. 2.5).

Then, by comparing the stress distribution in the medial and lateral compartments of the human knee joint, the peak contact stress and maximum shear stress in the articular cartilage increased by 200% more after a lateral meniscectomy than they did after a medial meniscectomy. These findings could partly explain the higher cartilage degeneration observed after a lateral meniscectomy. Another noteworthy result was that patterns of damage in a total meniscectomy model have better agreement with the clinical results when using relative increases in shear stress, rather than an absolute shear stress criterion.

A further FEA study (Peña et al. 2007) explored the influence of osteochondral defect size and location on the stress and strain concentrations around the defect rim. From those results, it was revealed that no stress concentration appeared

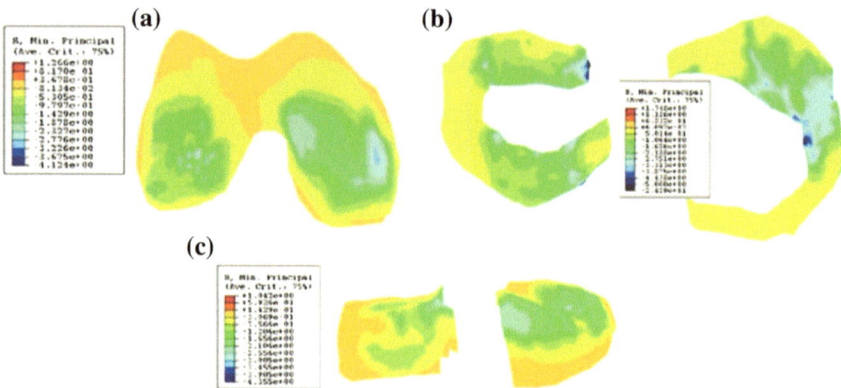

Fig. 2.4 Maximal compressive stress in the healthy tibio-femoral joint: **a** femoral cartilage, **b** menisci, **c** tibial cartilage (Peña et al. 2005)

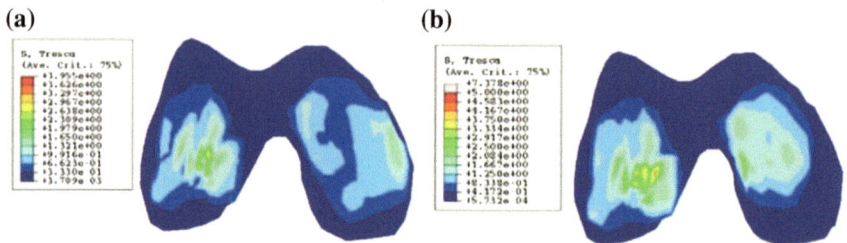

Fig. 2.5 Tresca stress distribution in the articular cartilage: **a** healthy joint and **b** total medial meniscectomy (Peña et al. 2005)

around the rim of small defects, since the stress distribution was mainly guarded by the meniscus contact. Nevertheless, significant stress concentration was found around the rim of large osteochondral defects, showing that this variance of the stress distribution has important clinical implications regarding the long-term integrity of the adjacent cartilage of osteochondral defects.

Later, Bae et al. (2012) investigated the effects of meniscectomy on degenerative osteoarthritis by estimating the contact area and contact pressure representing the normal part of the contact stress in soft tissues of the knee joint. A 3D FEM of the human lower limb was developed in order to model and simulate four cases; namely, the intact meniscus, and the partial, subtotal, and total meniscectomies of the medial meniscus. A vertical load of 570 N was applied on the sacrum through the center of the L5S1 disc and the facet of the sacral horn. The x-direction displacements of the nodes on the symmetric surfaces of the sacrum and coxal bones with the nodes on the distal surfaces of the tibia and fibula were all fixed to prevent translation and rotation, as shown in Fig. 2.6.

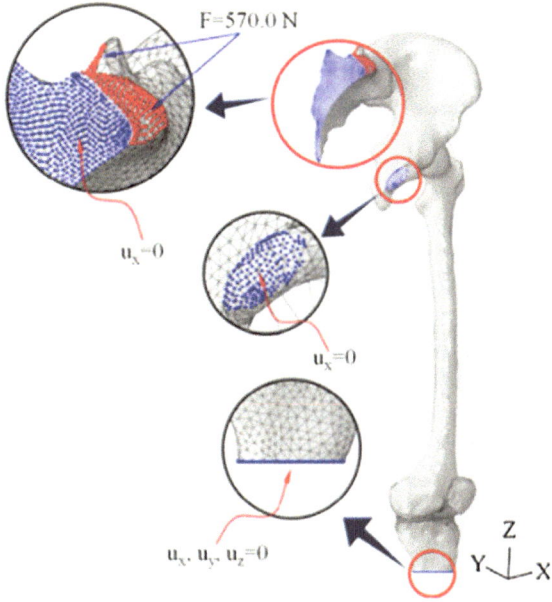

Fig. 2.6 Load and boundary conditions applied to the FEM of the lower limb (Bae et al. 2012)

As shown in Fig. 2.7, the cartilage-to-cartilage contact area in both the medial and lateral compartments increased significantly compared to the cartilage-to-meniscus contact area. Figure 2.8 indicates that the peak contact pressure in the meniscus increased significantly, but shows a slow increase in articular cartilage

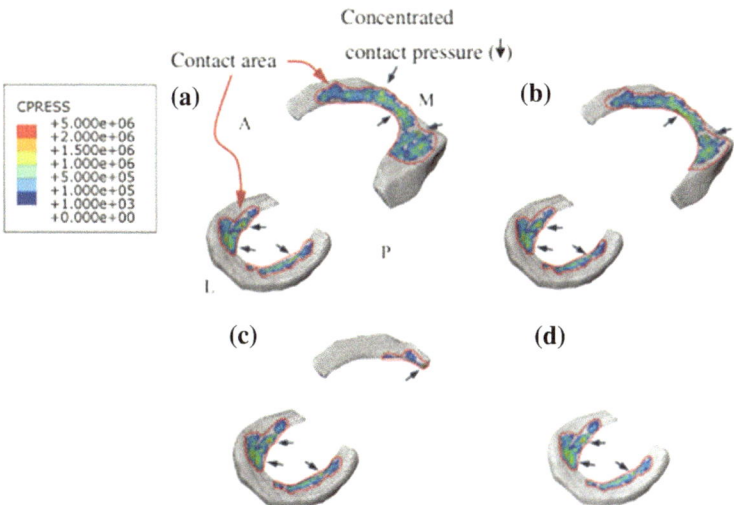

Fig. 2.7 Contact pressure distributions in **a** intact meniscus and **b–d** menisci under partial, sub-total, and total meniscectomy, respectively (Bae et al. 2012)

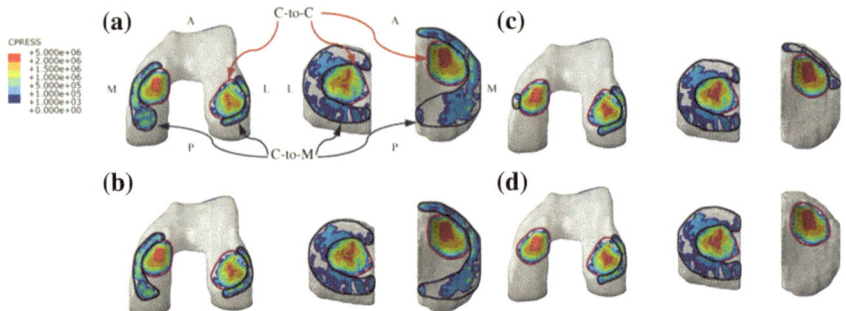

Fig. 2.8 Contact pressure distributions in the femoral and tibial articular cartilages following **a** no meniscectomy and **b–d** partial, sub-total, and total meniscectomy, respectively (Bae et al. 2012)

depending on the type of meniscectomy. Thus, the cartilage-to-cartilage contact area and the peak contact pressure in the meniscus were considered as significant parameters for evaluating degenerative OA based on the effects of contact area and contact stress on degenerative OA, which confirms the results reported by Fukubayashi and Kurosawa (1980) and Segal et al. (2009).

In summary, the results of this study revealed that partial meniscectomy may be selected as a better surgical treatment than sub-total or total meniscectomy, and a high possibility of degenerative OA is predicted after total meniscectomy. Moreover, medial meniscectomy likely gives rise to degenerative OA in both the medial and lateral compartments of a knee joint.

2.3 Current FEA Applications on Knee Joint Contact Analysis and Cartilage Disease

Tanska et al. (2015) created a 3D multi-scale model for the investigation of chondrocyte strains, stresses and fluid pressures during gait loading within healthy and medial meniscectomy knee joints. In this novel computational method, the joint level loading was coupled with the fibril-reinforced poroviscoelastic properties of the extracellular (ECM) and pericellular matrix (PCM) (Fig. 2.9). Additionally, the effect of a medial meniscectomy on chondrocyte responses was investigated.

As shown in Figs. 2.10 and 2.11, fluid pressures in the chondrocyte and cartilage tissue increased by about 30% in the meniscectomy joint compared to the normal, healthy joint. The elevated level of fluid pressure was detected during the entire stance phase of gait. Indeed, the medial meniscectomy substantially changed the maximum principal strains in the chondrocyte up to 60% compared to those in the ECM or PCM, while it did not affect chondrocyte volume or morphology. Results also suggested that during walking, chondrocyte deformations are not substantially altered due to a medial meniscectomy, while abnormal joint loading exposes

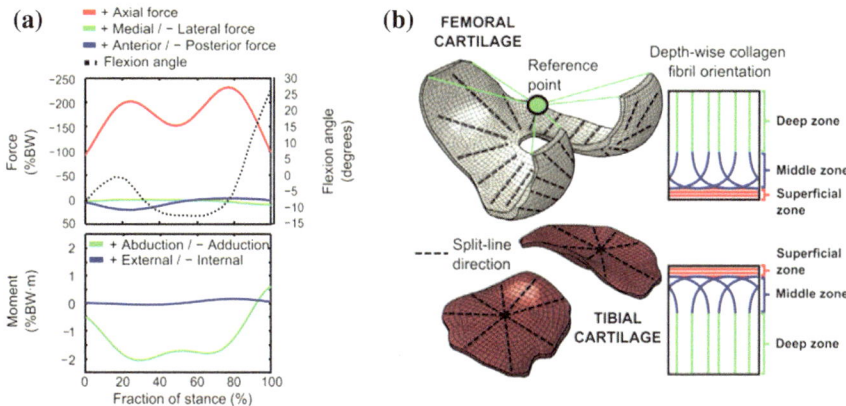

Fig. 2.9 Literature gait cycle input (**a**) and fibril-reinforced cartilage properties (**b**) (Tanska et al. 2015)

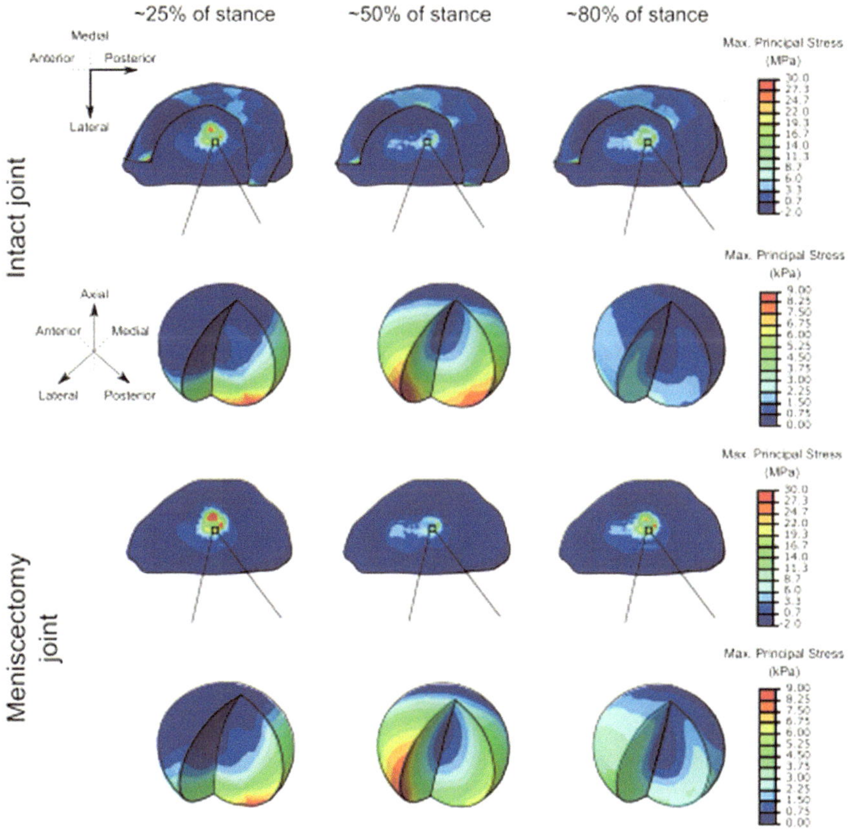

Fig. 2.10 Maximum principal stress of the medial compartment and chondrocyte of the knee joint in the intact and meniscectomy joint models (Tanska et al. 2015)

Fig. 2.11 Average fluid
pressure of the chondrocyte
(cell), PCM and ECM in the
intact and meniscectomy joint
models (on the top), and the
difference in average fluid
pressure between the models
as a function of the stance
(on the bottom)
(Tanska et al. 2015)

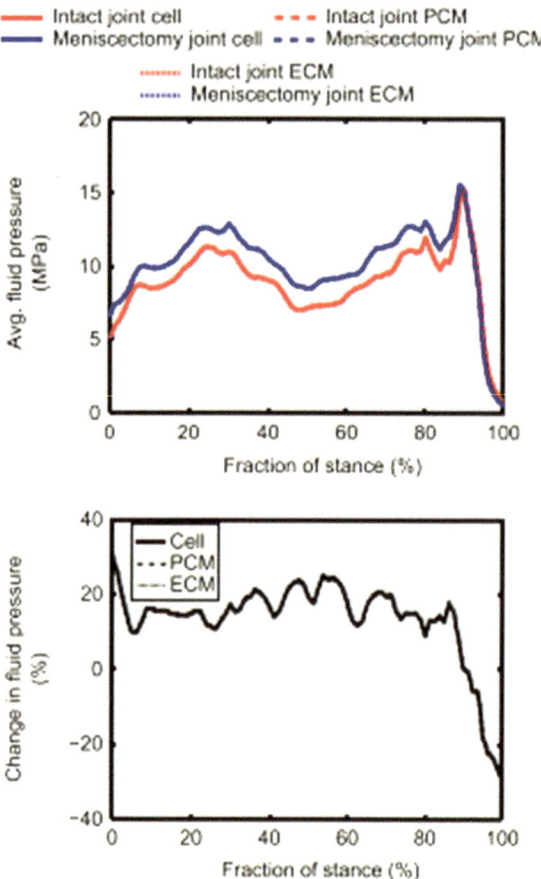

chondrocytes to elevated levels of fluid pressure and maximum principal strains. These increases subjected cells to more rapid and excessive changes in maximum principal strains, which may change cellular function and initiate OA.

Mononen et al. (2012) developed a 3D FEM of a healthy human knee joint based on an MRI, for the purpose of investigating the effect of superficial collagen fibril patterns and collagen fibrillation of cartilage on stresses and strains within a knee joint under axial impact loading of 1000 N.

The main results showed that the split-line patterns of femoral cartilage control stress and strain values and distributions in the knee joint, while the role of the split-lines is minor in distributing contact and pore pressures. Moreover, simulated medial OA increased tissue strains in both the medial and lateral femoral condyles with an increase in the contact and pore pressures in the lateral femoral condyle.

As shown in Fig. 2.12, the first row showed the local strain distributions in the anterior–posterior (E1) directions and the second row in the medial-lateral (E3) directions, while the third and fourth rows showed maximum and minimum

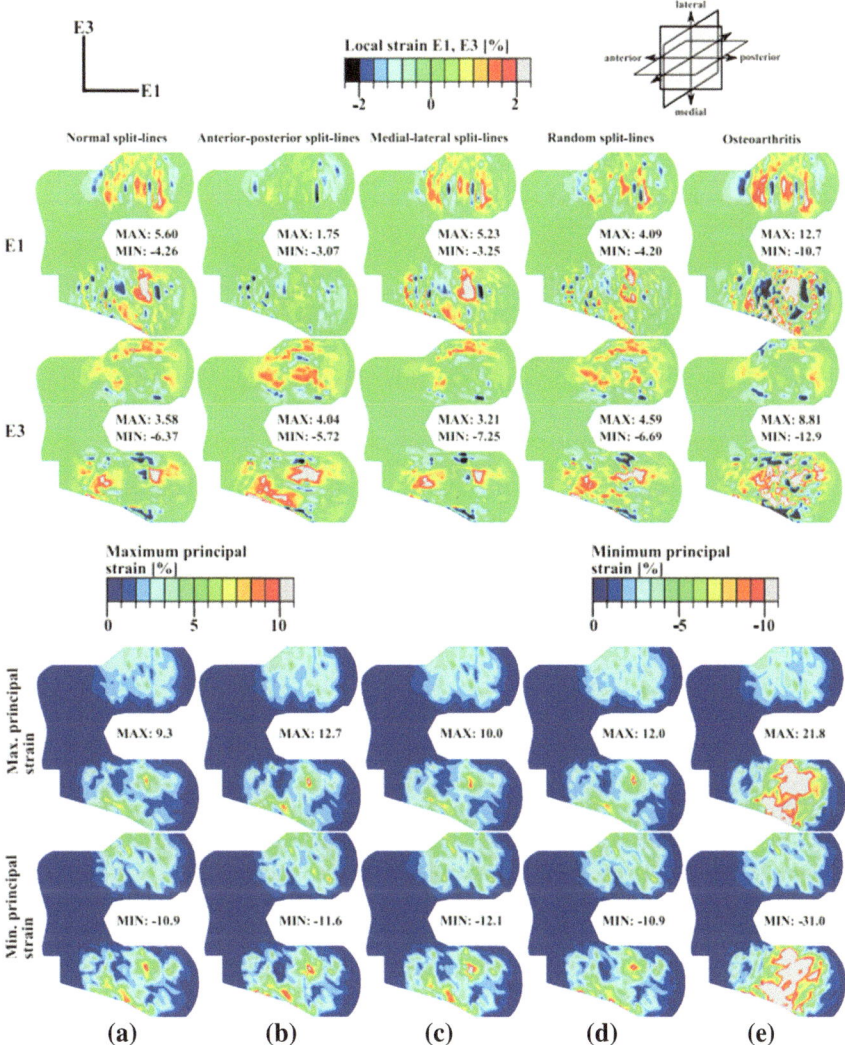

Fig. 2.12 Effect of the femoral split-line patterns and OA changes on the local strains, as well as maximum and minimum principal strains on the surface of femoral cartilage (Mononen et al. 2012)

principal strain distributions, respectively. Each column showed the effect of different split-line patterns on the local, maximum principal and minimum principal strain distributions in the following order: normal split-line pattern (a); anterior posterior split-line pattern (b); medial–lateral split-line pattern (c); random split-line pattern (d); and osteoarthritis in the medial femoral cartilage (e).

To conclude, this study provides further evidence of the importance of collagen fibril organization for the optimal function of articular cartilage, suggesting that OA

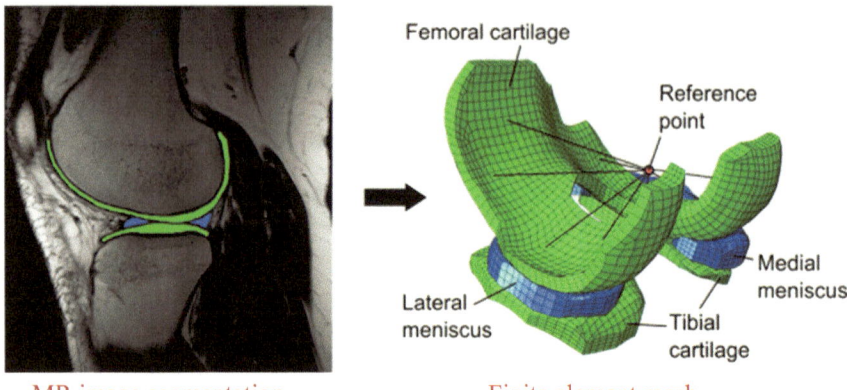

MR image segmentation   Finite element mesh

Fig. 2.13 3D FEM of the human knee joint (Halonen et al. 2013)

in the medial femoral cartilage represents a possible risk location in the lateral femoral condyle that could ultimately promote progression of OA.

Later, Halonen et al. (2013) characterized the effects of collagen orientation, collagen distribution and proteoglycan distribution on knee joint stresses and strains. Thereby, a 3D FEM of the human knee joint (Fig. 2.13) was created using five fibril-reinforced poroviscoelastic models with different depth-wise tissue structure.

For each model, strains and stresses were evaluated at four different depths in the medial tibial compartment during a gait cycle and at mechanical equilibrium.

The results of this work made it known that the depth-wise proteoglycan gradient substantially increased stresses and axial strains in the superficial layer, but reduced stresses and strains in the deep layer under dynamic and static loading of the joint. Additionally, the fibril volume density distribution had a minor effect on the mechanical behavior of the tibial articular cartilage in the knee joint during both dynamic joint loading and at mechanical equilibrium (Fig. 2.14). Finally, the results of this study suggested that a knee joint model, which incorporated the arcade-like collagen orientation in cartilage, provided an acceptable approach for analyzing loading patterns in cartilage during the gait cycle. However, the depth-wise PG distribution had an important influence in the simulations of equilibrium loading.

Considering the important role of menisci in knee joint function, several FE models of the knee joint that include meniscus-cartilage contact provide a better understanding of knee mechanics and tissue interactions through computational models of the tibio-menisco-femoral structure. In this context, further study of Guess et al. (2010) created dynamic 3D anatomical knee models with a computational efficiency sufficient for incorporation into neuromusculoskeletal models within a multibody (MB) framework. First, FE models of the subject knee included isolated models of the lateral meniscus, and medial meniscus (MEN models) were developed to generate force-displacement solution sets for the determination of MB

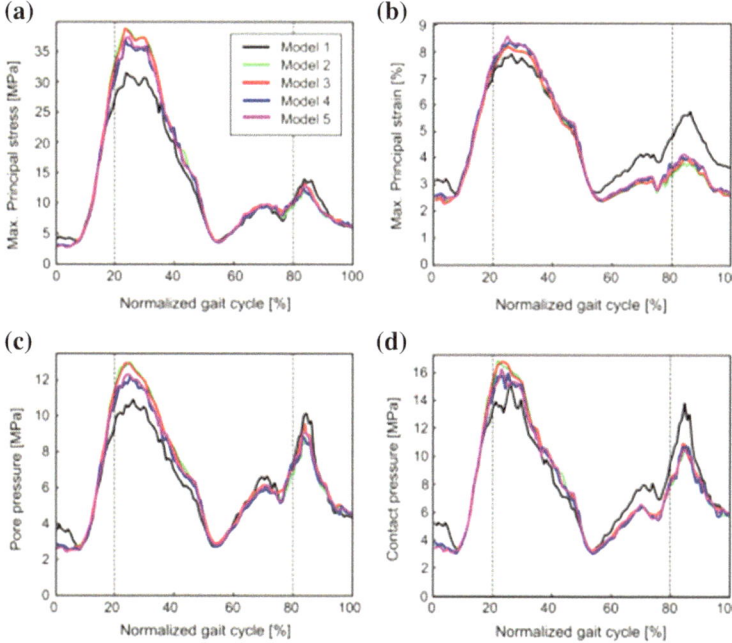

Fig. 2.14 Maximum principal stresses (**a**), maximum principal strains (**b**), pore pressures (**c**) and contact pressures (**d**) at the contact surface of the medial tibiofemoral joint cartilage during the gait cycle. Dashed lines indicate the peak forces of the gait cycle (Halonen et al. 2013)

meniscus spring matrix parameters. Then, the Tibio-Femoral (TF) model was developed to generate compressive force-displacement solution sets for the determination of MB-compliant contact parameters for cartilage-to-cartilage contact. Finally, The Tibio-Menisco-Femoral (TMF) joint model was developed to validate the FE model with an existing dataset from the literature (Donahue et al. 2002) and to provide force-displacement solution sets for the determination of MB-compliant contact parameters for meniscus-to-cartilage contact.

As shown in Fig. 2.15, the same compressive load of 800 N was applied at the distal tibia during tibio-femoral simulations for both the FE and MB TMF models.

Results culled from this study revealed that RMS errors between finite element displacement and MB displacement after parameter optimization were 0.017 mm for the lateral meniscus and 0.051 mm for the medial meniscus. RMS errors between model-predicted and experimental cadaver kinematics during a walk cycle were less than 11 mm in translation and less than 7° in orientation. Then, the addition of the menisci to the model significantly reduced contact between the tibia cartilage and femur cartilage for the lateral side, while it was not statistically significant on the medial side (Fig. 2.16). In addition, Fig. 2.17 shows that the highest contact force for a single element on the lateral side was about twice that on the medial side. Moreover, the computational time for the MB TMF model was

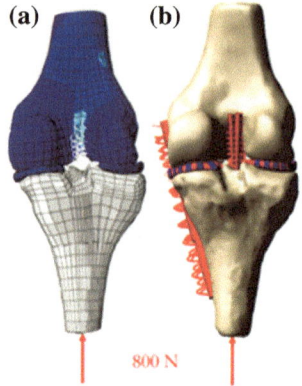

Fig. 2.15 Finite element (**a**) and multibody (**b**) models of the tibio-femoral joint with menisci (Guess et al. 2010)

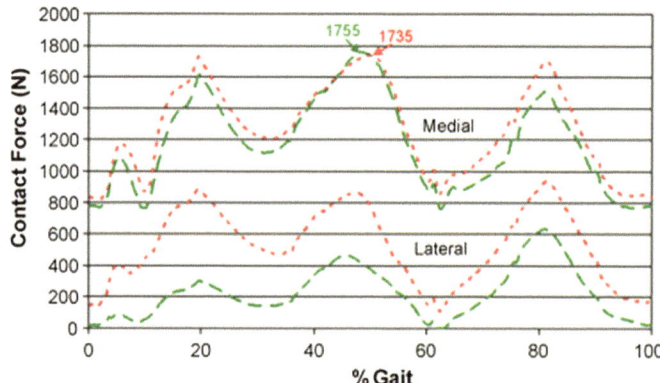

Fig. 2.16 Predicted tibio-femoral contact forces for a model with menisci (**1**) and without menisci (**2**) over one walk cycle (Guess et al. 2010)

30 s of computation time for each second of simulation time, while the FE TMF model required 128 s of computation time for each second of simulation time.

Later, Adouni et al. (2012) developed a validated FE model of a healthy knee joint with the aim of computing muscle forces and determining tissues stresses and strains during the gait cycle. In vivo kinematics–kinetics data and ground reaction forces in asymptomatic subjects were reported in the FE model (Fig. 2.18), which includes the bony structure (tibia, patella and femur), the major tibio-femoral (ACL, PCL, LCL and MCL) and patello-femoral (MPFL and LPFL) ligaments, as well as the patellar tendon (PT). The quadriceps components are the vastus medialis obliqus (VMO), rectus femoris (RF), vastus intermidus medialis (VIM), and vastus lateralis (VL). The hamstring components include the biceps femoris (BF) and semimembranous (SM). The gastrocnemius components are the gastrocnemius

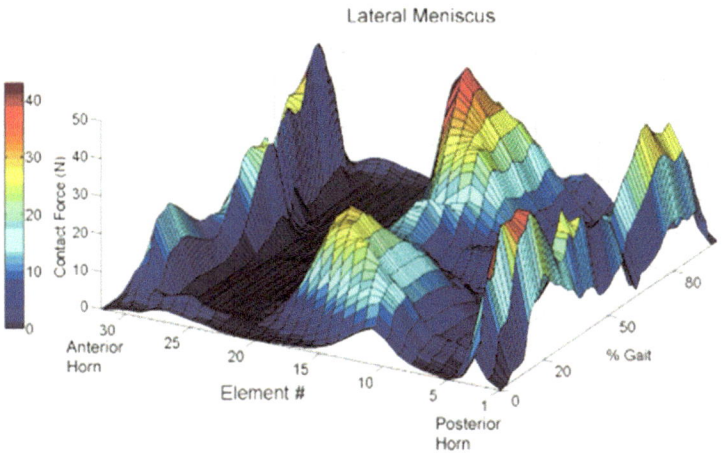

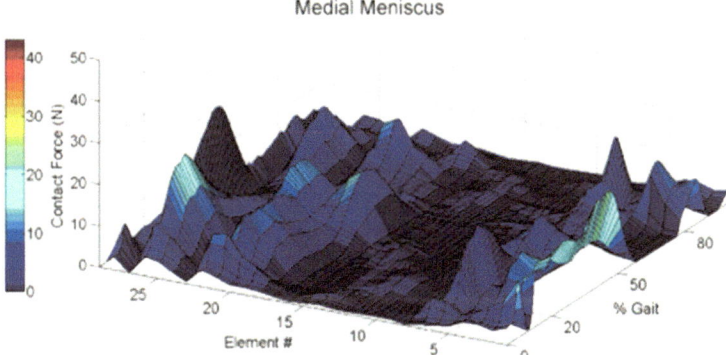

Fig. 2.17 Force predicted by the compliant contact model between each meniscus element and the tibia during a walk cycle (Guess et al. 2010)

medial (GM) and gastrocnemius lateral (GL), the TRIPOD made of Sartorius (SR), gracilis (GR) and semitendinosus (ST).

In this case, the muscle components are modeled by uniaxial elements with orientations at full extension taken from the literature (Sakai et al. 1996; Aalbersberg et al. 2005; Delp et al. 2007).

Results revealed that muscle forces and joint response are substantially altered during the stance phase, and that ligament forces may be accurately computed through a simplified joint model. Then, performing the analyses at 6 periods from beginning to the end (0, 5, 25, 50, 75 and 100%), the hamstring forces peaked at

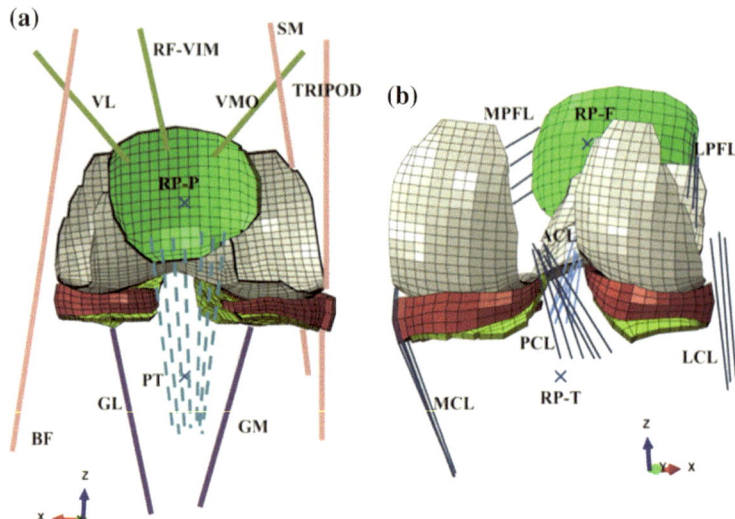

Fig. 2.18 Knee joint FE model: **a** anterior view, **b** posterior view (Adouni et al. 2012)

5%, and the quadriceps forces at 25%, whereas the gastrocnemius forces peaked at 75%. The ACL Force reached its maximum of 343 N at 25% and decreased thereafter.

As shown in Fig. 2.19, contact stresses, as well as cartilage stresses and strains, are influenced by compartmental loads. The cartilage on the medial compartment experiences greater compressive stresses and strains, as well as a posterior shift in the contact area. Moreover, contact forces reached maximum at the 5, 25 and 75% periods, with the medial compartment carrying a major portion of the load and experiencing larger relative movements and cartilage strains. Much smaller contact stresses were computed at the patellofemoral joint.

Other literature researchers aim to study the mechanical behavior of the patellofemoral (PF) joint of the human knee joint, given its importance to optimal knee function. Ali et al. (2016) developed a 3D FE model of the PF joint (Fig. 2.20) to predict PF kinematics, contact mechanics, quadriceps force, patellar tendon moment arm and patellar tendon angle of the cruciate resected conditions. Measurements from a series of dynamic in vitro cadaveric experiments were employed to create FE models of the knee for three specimens.

Model findings for the intact and cruciate resected trials successfully matched experimental kinematics and peak quadriceps forces. Cruciate resections either demonstrated increased patellar tendon loads or increased joint reaction forces. As shown in Fig. 2.21, PF contact distributions displayed superior travel capabilities and the contact area increased as a function of knee flexion. Therefore, the models developed in this work can be employed to simulate soft-tissue injury and repair and used to quantitatively assess the effect of surgical decisions during ACL or PCL reconstruction on PF mechanics and extensor mechanism efficiency and function.

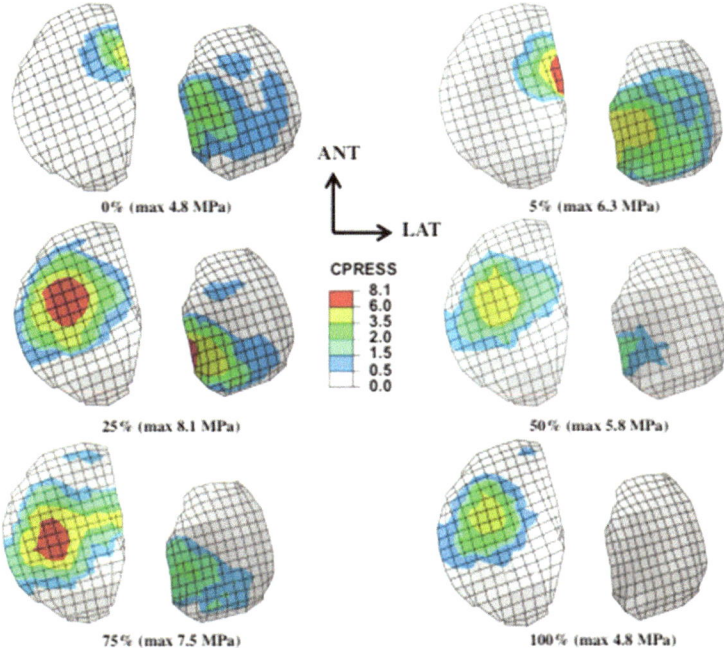

Fig. 2.19 Predicted contact pressure at the articular surface of the lateral and medial tibial plateaus at different stance phases (Adouni et al. 2012)

Overall, the use of accurate 3D FEM helps us to better understand the mechanical behavior of the different components of the knee joint. This approach can benefit many areas of medical development, such as design and the testing of orthopedic implants. Moreover, the comparison of the effectiveness of the different clinical treatments leads to improved surgical techniques. The FEA method mentioned above facilitates the ability to identify those people most susceptible to OA and develop preventive measures, and it can also be employed for long-term follow-up assessment of clinical treatment. Indeed, it addresses the need to study stress distribution in the knee cartilage based on subject-specific loading conditions and geometry. Since it is a relatively inexpensive method, and with the cost of MRIs becoming cheaper and computational biomechanics becoming faster with stronger computational abilities, it becomes a worthwhile tool that provides insightful understanding of the clinical setting to assist in subject-specific OA prevention methods such as HTO.

Within this framework, the study by Mootanah et al. (2014) sets one's sights on the validation of the capability of a subject-specific model for predicting joint forces and pressures experienced under different load conditions. An additional target is the assessment of the utility of a 3D computational knee model in obtaining improved outcomes, making it a subject-specific tool for guiding orthopedic surgeons towards obtaining realignment of a misaligned knee that minimises peak

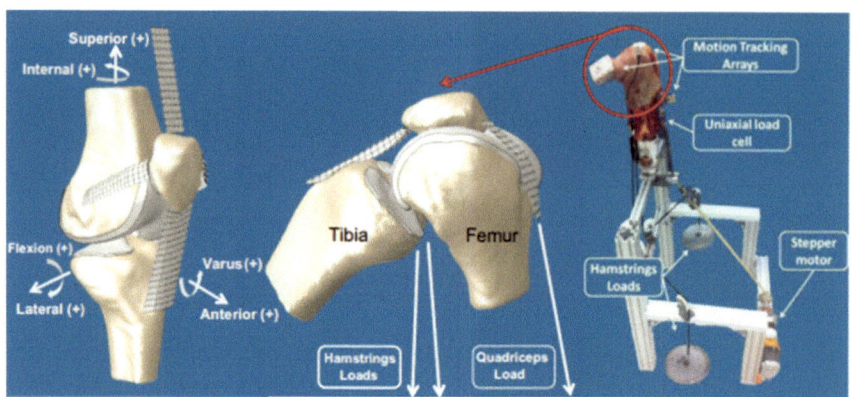

Fig. 2.20 Cadaver knee mounted in a muscle loading rig (MLR) (on the right) and its computational representation (on the left) (Ali et al. 2016)

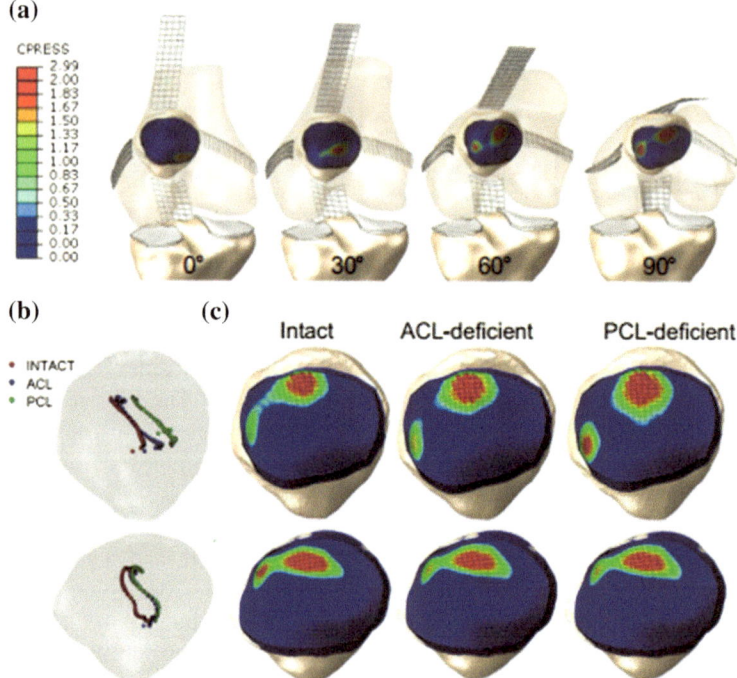

Fig. 2.21 **a** PF contact pressure distributions at different knee flexion angles, **b** the PF contact center of pressure through the flexion activity, and **c** contact distribution at $\sim 90°$ is shown for a representative specimen under intact and cruciate-deficient conditions (Ali et al. 2016)

pressure and equilibrates stress distribution within the joint. At that time, a 3D FEM was developed and validated experimentally to predict knee joint contact forces and pressures for different degrees of malalignment. In order to replicate in vitro testing, the proximal femur was mechanically grounded in all degrees of freedom. The distal tibia was free in five degrees of freedom and fixed in 20° of sagittal plane knee flexion to simulate the end of weight acceptance during the stance phase of gait, when the knee joint is subject to higher loading. Then, an axial load of 374 N was applied along the tibia and varus/valgus bending moments, ranging from 0 to 15 Nm, were applied about the knee joint's center.

By comparing the intra-articular force and pressure measurements (Fig. 2.22), results revealed that the percentage of full scale error between FE-predicted and in vitro-measured values in the medial and lateral compartments were 6.67 and 5.94%, respectively, for normalized peak pressure values, and 7.56 and 4.48%, respectively, for normalized force values. Accordingly, the FE knee joint model designed in this study was considered capable of predicting normalized intra-articular

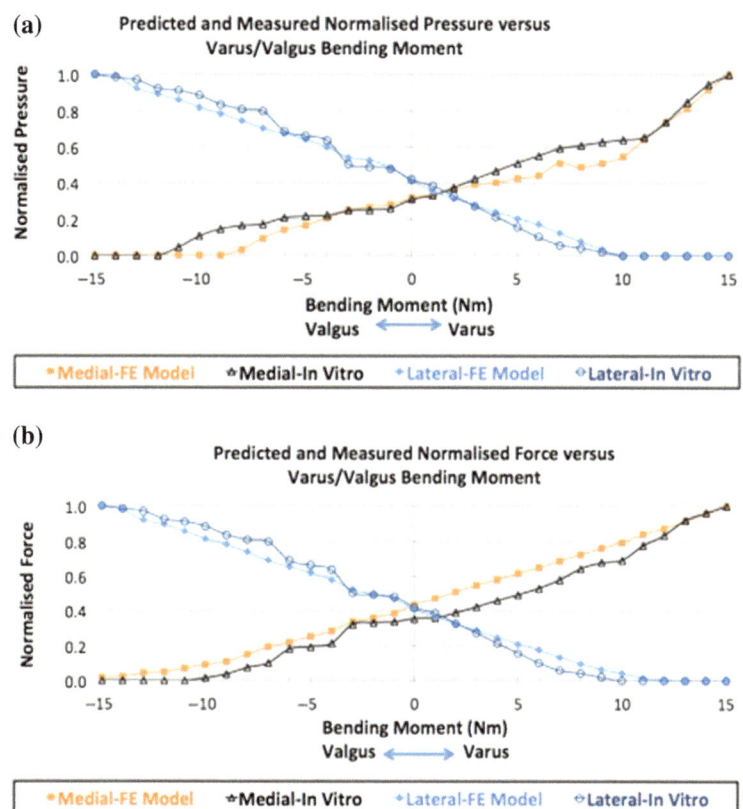

Fig. 2.22 In vitro and FE-predicted medial and lateral compartment loading for normalised peak pressure (**a**) and normalised force (**b**) (Mootanah et al. 2014)

Fig. 2.23 Finite element knee joint model: **a** kneeling position and **b** standing position (Wang et al. 2014)

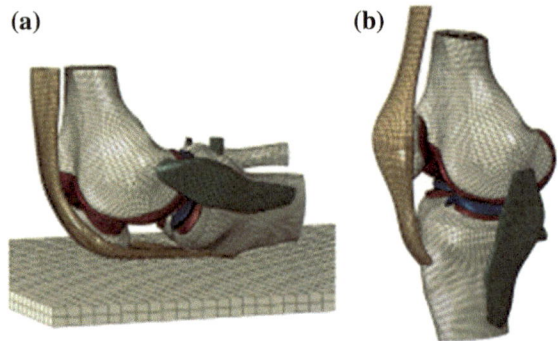

pressure and forces for different loading conditions. Subsequently, it will yield further improvements in predicting alignment correction and clinical outcomes.

In their study, Wang et al. (2014) compared the stress distributions on knee joint cartilage between the kneeling and standing positions. Within this aim, two 3D FE models of the knee joint for both postures were developed (Fig. 2.23). Identical material properties were assigned to the models. There was a compulsory reminder that the cartilage was assumed to be isotropic elastic with an elastic modulus of 10 MPa and Poisson's ratio ranging from 0.05 to 0.45.

In order to simulate the physiological loading of the knee joint, the femur was fixed in space, and the tibia set free in all degrees of freedom. Stresses in the knee joint components were calculated under a compressive load of up to 1000 N related with muscle forces, such as quadriceps 215 N, biceps 31 N, and semimembranous 54 N. The loads and boundary conditions were kept the same in the two models, except for the tibia in the standing model, where the compressive load was applied in the rigid plane tied to the end surface of the tibia.

As shown in Fig. 2.24, a clear difference is found in high-stress regions between kneeling and standing. Both the peak von Mises stress and contact pressure on the cartilage were larger in kneeling. During kneeling, the contact pressure reached 4.25 MPa under a 300 N compressive load. It then increased to 4.66 MPa at 600 N and 5.15 MPa at 1000 N. Moreover, the findings of this study revealed that changing the Poisson's ratio of the cartilage, which represents changes in compressibility caused by different loading rates, was found to have an influence on the magnitude of stress.

To conclude, this work has proven that with the same magnitude of compressive loads, kneeling can result in greater stress on the cartilage and produce quite a different stress distribution when compared to normal standing. The higher stress and shifted loading area underlined the risk of knee disorders, which lead to the progression of OA.

On the other hand, Smith et al. (2016) explored the influence of joint alignment and uncertain ligament properties on total knee replacement (TKR) loading during walking. A subject-specific knee model was created that included deformable contact, ligamentous structures, and six degrees-of-freedom (DOF) tibiofemoral and

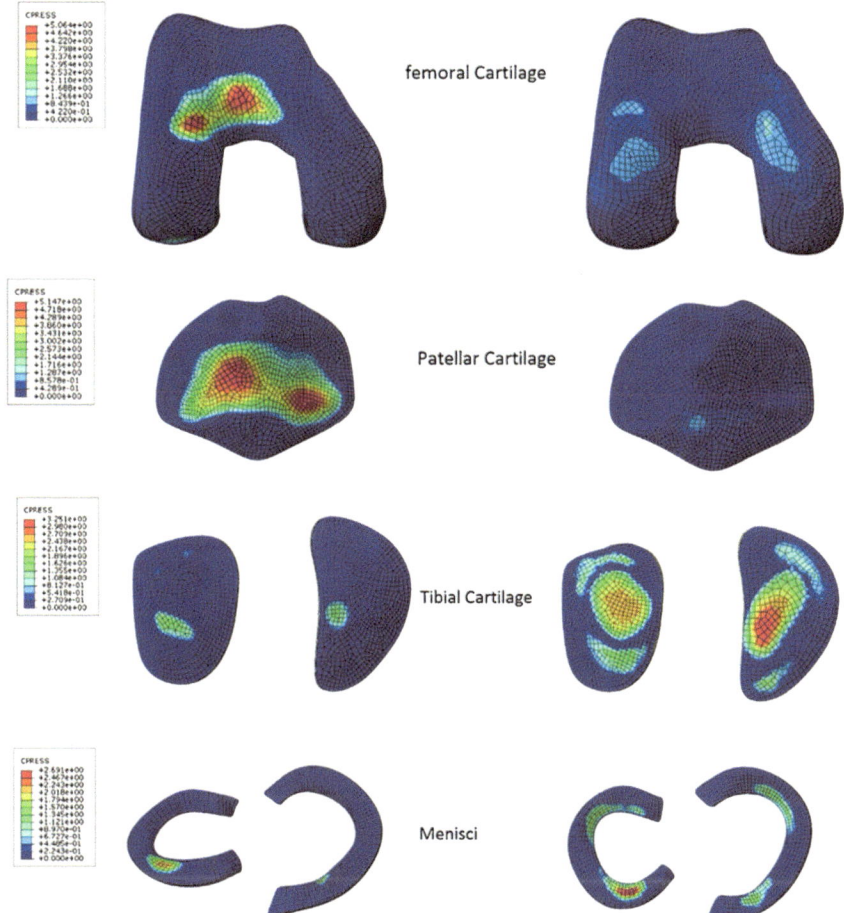

Fig. 2.24 Contact pressure on the cartilage of a knee joint under 1000 N compressive loads with Poisson's ratio set to 0.45: **a** kneeling position and **b** standing position (Wang et al. 2014)

patellofemoral joints. In order to achieve accurate results, a novel numerical optimization technique was used to simultaneously predict muscle forces, secondary knee kinematics, ligament forces, and joint contact pressures from standard gait analysis data collected on the subject. Also, a Monte Carlo approach was used to assess the effect of uncertainties in ligament stiffness and reference strains on both ligament forces and tibiofemoral contact force predictions.

Significant agreement between the nominal knee model predictions of medial, lateral, and total contact forces during gait and TKR measures was noticed, with root-mean-square (rms) errors of 0.23, 0.22, and 0.33 body weight (BW), respectively. More than that, Coronal plane component alignment did not affect total knee contact loads, but did modify the medial lateral load distribution (Fig. 2.25),

Fig. 2.25 Sensitivity of the joint contact forces to variations in coronal plane component alignment for smooth gait (Smith et al. 2016)

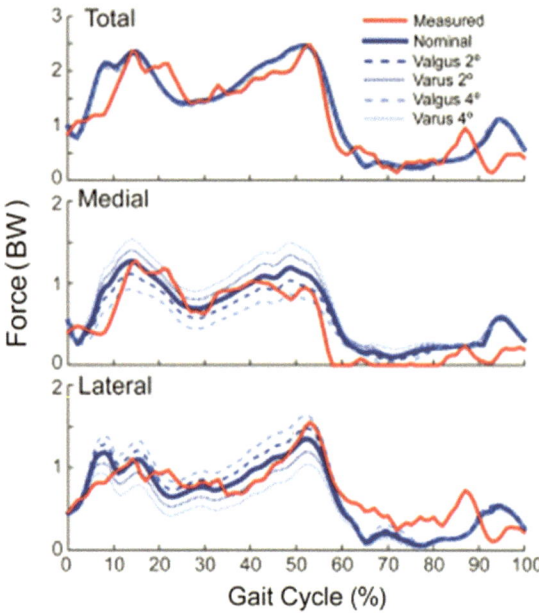

with 4° varus/valgus rotations in component alignment inducing +17% and −23% changes in the first peak medial tibiofemoral contact forces, respectively. Also, ligament properties had substantial influence on the TKR load distributions, with the medial collateral ligament and iliotibial band (ITB) properties having the largest effects on medial and lateral compartment loading, respectively.

Therefore, the proposed framework provides a realistic objective approach for modeling TKR components virtually, considering parametric uncertainty and predicting the effects of joint alignment and soft tissue balancing procedures on TKR function in movement.

In order to evaluate biomechanical interaction between the meniscus and the cartilage in medial compartment knee OA, Łuczkiewicz et al. (2016) developed three knee models on the basis of knee geometry from the Open Knee project. Validation of these models was based on date from the literature (Burkhart et al. 2013). It was assumed that the articular cartilage's stiffness has small impact on meniscus biomechanics (Yao et al. 2006; Meakin et al. 2003). The thickness of medial cartilages in the intact knee model (model 1) was reduced by approximately 50% to obtain a medial knee OA model. Two variants of a medial knee OA model with congruent (model 2) and incongruent (model 3) contact surfaces were analyzed to study the influence of congruency. A nonlinear static analysis for one compressive load of 1000 N was performed to investigate the influence of cartilage thickness on the ability of the menisci to distribute load.

Results from this study revealed that the variation in articular cartilage geometry leads to significant reduction of the role of the meniscus in load transmission and

Fig. 2.26 Comparison of
principal compressive stresses
on the tibial cartilages for the
three models (Łuczkiewicz
et al. 2016)

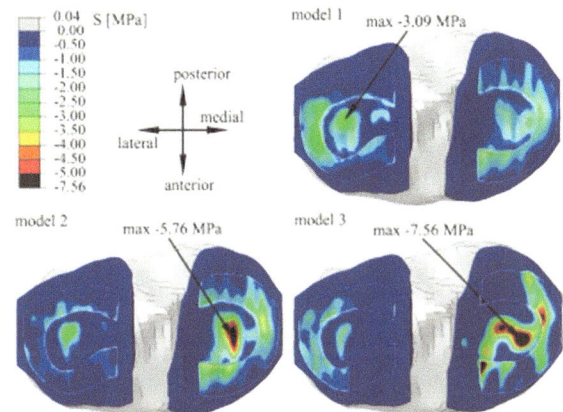

the contact area between the meniscus and the cartilage. Figure 2.26 shows the
maximal compressive stress distribution found in the tibial plateau. In the model
with incongruent contact surfaces, the value of medial meniscus external shift was
95.3% greater, while the contact area between the tibial cartilage and medial
meniscus was 50% lower than in the congruent contact surfaces model. The medial
meniscus carried only 48.4% of load in the medial compartment, in comparison to
71.2% in the healthy knee model, after the non-uniform reduction of cartilage
thickness. Accordingly, medial knee OA may increase the risk of meniscal extru-
sion in the medial compartment of the knee joint.

However, this work lacks a physical validation for the knee joint geometry from
the Open Knee project. Another limitation is the assumption of the same material
properties for cartilages and menisci in the healthy knee model and the knee models
with medial compartment OA.

References

Aalbersberg, S., et al. (2005). Orientation of tendons in vivo with active and passive knee muscles.
   Journal of Biomechanics, 38(9), 1780–1788.
Adouni, M., Shirazi-Adl, A., & Shirazi, R. (2012). Computational biodynamics of human knee
   joint in gait: From muscle forces to cartilage stresses. *Journal of Biomechanics, 45*(12),
   2149–2156.
Ali, A. A., et al. (2016). Validation of predicted patellofemoral mechanics in a finite element
   model of the healthy and cruciate-deficient knee. *Journal of Biomechanics, 49*(2), 302–309.
Bae, J. Y., et al. (2012). Biomechanical analysis of the effects of medial meniscectomy on
   degenerative osteoarthritis. *Medical & Biological Engineering & Computing, 50*(1), 53–60.
Baldwin, M. A., et al. (2012). Dynamic finite element knee simulation for evaluation of knee
   replacement mechanics. *Journal of Biomechanics, 45*(3), 474–483.
Beillas, P., et al. (2004). A new method to investigate in vivo knee behavior using a finite element
   model of the lower limb. *Journal of Biomechanics, 37*(7), 1019–1030.

Bendjaballah, M. Z., Shirazi-Adl, A., & Zukor, D. J. (1997). Finite element analysis of human knee joint in varus-valgus. *Clinical Biomechanics, 12*(3), 139–148.

Burkhart, T. A., Andrews, D. M., & Dunning, C. E. (2013). Finite element modeling mesh quality, energy balance and validation methods: A review with recommendations associated with the modeling of bone tissue. *Journal of Biomechanics, 46*(9), 1477–1488.

Delp, S. L., et al. (2007). OpenSim: Open-source software to create and analyze dynamic simulations of movement. *IEEE Transactions on Biomedical Engineering, 54*(11), 1940–1950.

Dhaher, Y. Y., Kwon, T.-H., & Barry, M. (2010). The effect of connective tissue material uncertainties on knee joint mechanics under isolated loading conditions. *Journal of Biomechanics, 43*(16), 3118–3125.

Donahue, T. L. H., et al. (2002). A finite element model of the human knee joint for the study of tibio-femoral contact. *Journal of Biomechanical Engineering, 124*(3), 273–280.

Ellis, B. J., et al. (2006). Medial collateral ligament insertion site and contact forces in the ACL-deficient knee. *Journal of Orthopaedic Research, 24*(4), 800–810.

Erdemir, A. (2013). Open knee: A pathway to community driven modeling and simulation in joint biomechanics. *Journal of Medical Devices, 7*(4), 40910.

Erdemir, A. (2016). Open knee: Open source modeling and simulation in knee biomechanics. *Journal of Knee Surgery, 29*(2), 107–116.

Fernandes, D. J. C. (2014). Finite element analysis of the ACL-deficient Knee.

Fukubayashi, T., & Kurosawa, H. (1980). The contact area and pressure distribution pattern of the knee: A study of normal and osteoarthrotic knee joints. *Acta Orthopaedica Scandinavica, 51* (1–6), 871–879.

Gardiner, J. C., & Weiss, J. A. (2003). Subject-specific finite element analysis of the human medial collateral ligament during valgus knee loading. *Journal of Orthopaedic Research, 21*(6), 1098–1106.

Gasser, T. C., Ogden, R. W., & Holzapfel, G. A. (2006). Hyperelastic modelling of arterial layers with distributed collagen fibre orientations. *Journal of the Royal Society, Interface, 3*(6), 15–35.

Godest, A. C., et al. (2002). Simulation of a knee joint replacement during a gait cycle using explicit finite element analysis. *Journal of Biomechanics, 35*(2), 267–275.

Guess, T. M., et al. (2010). A subject specific multibody model of the knee with menisci. *Medical Engineering & Physics, 32*(5), 505–515.

Halloran, J. P., Petrella, A. J., & Rullkoetter, P. J. (2005). Explicit finite element modeling of total knee replacement mechanics. *Journal of Biomechanics, 38*(2), 323–331.

Halloran, J. P., et al. (2010). Concurrent musculoskeletal dynamics and finite element analysis predicts altered gait patterns to reduce foot tissue loading. *Journal of Biomechanics, 43*(14), 2810–2815.

Halonen, K. S., et al. (2013). Importance of depth-wise distribution of collagen and proteoglycans in articular cartilage—A 3D finite element study of stresses and strains in human knee joint. *Journal of Biomechanics, 46*(6), 1184–1192.

Hirokawa, S., & Tsuruno, R. (2000). Three-dimensional deformation and stress distribution in an analytical/computational model of the anterior cruciate ligament. *Journal of Biomechanics, 33* (9), 1069–1077.

Kiapour, A. M., et al. (2014). The effect of ligament modeling technique on knee joint kinematics: a finite element study. *Applied mathematics, 4*(5A), 91.

Li, G., Suggs, J., & Gill, T. (2002). The effect of anterior cruciate ligament injury on knee joint function under a simulated muscle load: A three-dimensional computational simulation. *Annals of Biomedical Engineering, 30*(5), 713–720.

Li, G., et al. (1999). A validated three-dimensional computational model of a human knee joint. *Journal of Biomechanical Engineering, 121*(6), 657–662.

Limbert, G., Middleton, J., & Taylor, M. (2004). Finite element analysis of the human ACL subjected to passive anterior tibial loads. *Computer Methods in Biomechanics and Biomedical Engineering, 7*(1), 1–8.

Łuczkiewicz, P., et al. (2016). The influence of articular cartilage thickness reduction on meniscus biomechanics. *PLoS One, 11*(12), e0167733.

Marlow, R. S. (2003). A general first-invariant hyperelastic constitutive model. *Constitutive Models for Rubber*, 157–160.

Meakin, J. R., et al. (2003). Finite element analysis of the meniscus: The influence of geometry and material properties on its behaviour. *The Knee, 10*(1), 33–41.

Moglo, K. E., & Shirazi-Adl, A. (2003). Biomechanics of passive knee joint in drawer: Load transmission in intact and ACL-deficient joints. *The Knee, 10*(3), 265–276.

Mononen, M. E., et al. (2012). Effect of superficial collagen patterns and fibrillation of femoral articular cartilage on knee joint mechanics—A 3D finite element analysis. *Journal of Biomechanics, 45*(3), 579–587.

Mootanah, R., et al. (2014). Development and validation of a computational model of the knee joint for the evaluation of surgical treatments for osteoarthritis. *Computer Methods in Biomechanics and Biomedical Engineering, 17*(13), 1502–1517.

Park, H.-S., et al. (2010). A knee-specific finite element analysis of the human anterior cruciate ligament impingement against the femoral intercondylar notch. *Journal of Biomechanics, 43* (10), 2039–2042.

Peña, E., Calvo, B., et al. (2005). Finite element analysis of the effect of meniscal tears and meniscectomies on human knee biomechanics. *Clinical Biomechanics, 20*(5), 498–507.

Peña, E., et al. (2006a). A three-dimensional finite element analysis of the combined behavior of ligaments and menisci in the healthy human knee joint. *Journal of Biomechanics, 39*(9), 1686–1701.

Peña, E., et al. (2006b). Why lateral meniscectomy is more dangerous than medial meniscectomy. A finite element study. *Journal of Orthopaedic Research, 24*(5), 1001–1010.

Peña, E., et al. (2007). Effect of the size and location of osteochondral defects in degenerative arthritis. A finite element simulation. *Computers in Biology and Medicine, 37*(3), 376–387.

Peña, E., et al. (2008). Computer simulation of damage on distal femoral articular cartilage after meniscectomies. *Computers in Biology and Medicine, 38*(1), 69–81.

Penrose, J. M. T., et al. (2002). Development of an accurate three-dimensional finite element knee model. *Computer Methods in Biomechanics & Biomedical Engineering, 5*(4), 291–300.

Pioletti, D. P., et al. (1998). Viscoelastic constitutive law in large deformations: Application to human knee ligaments and tendons. *Journal of Biomechanics, 31*(8), 753–757.

Ramaniraka, N. A., Terrier, A., et al. (2005). Effects of the posterior cruciate ligament reconstruction on the biomechanics of the knee joint: a finite element analysis. *Clinical Biomechanics, 20*(4), 434–442.

Ramaniraka, N. A., et al. (2007). Biomechanical evaluation of intra-articular and extra-articular procedures in anterior cruciate ligament reconstruction: a finite element analysis. *Clinical Biomechanics, 22*(3), 336–343.

Sakai, N., et al. (1996). Quadriceps forces and patellar motion in the anatomical model of the patellofemoral joint. *The Knee, 3*(1–2), 1–7.

Sathasivam, S., & Walker, P. S. (1997). A computer model with surface friction for the prediction of total knee kinematics. *Journal of Biomechanics, 30*(2), 177–184.

Segal, N. A., et al. (2009). Baseline articular contact stress levels predict incident symptomatic knee osteoarthritis development in the MOST cohort. *Journal of Orthopaedic Research, 27* (12), 1562–1568.

Smith, C. R., et al. (2016). The influence of component alignment and ligament properties on tibiofemoral contact forces in total knee replacement. *Journal of Biomechanical Engineering, 138*(2), 21017.

Song, Y., et al. (2004). A three-dimensional finite element model of the human anterior cruciate ligament: A computational analysis with experimental validation. *Journal of Biomechanics, 37* (3), 383–390.

Tanska, P., Mononen, M. E., & Korhonen, R. K. (2015). A multi-scale finite element model for investigation of chondrocyte mechanics in normal and medial meniscectomy human knee joint during walking. *Journal of Biomechanics, 48*(8), 1397–1406.

Wan, C., Hao, Z., & Wen, S. (2013). The effect of the variation in ACL constitutive model on joint kinematics and biomechanics under different loads: A finite element study. *Journal of Biomechanical Engineering, 135*(4), 41002.

Wang, Y., Fan, Y., & Zhang, M. (2014). Comparison of stress on knee cartilage during kneeling and standing using finite element models. *Medical Engineering & Physics, 36*(4), 439–447.

Xie, F., et al. (2009). A study on construction three-dimensional nonlinear finite element model and stress distribution analysis of anterior cruciate ligament. *Journal of Biomechanical Engineering, 131*(12), 121007.

Yao, J., Funkenbusch, P. D., et al. (2006a). Sensitivities of medial meniscal motion and deformation to material properties of articular cartilage, meniscus and meniscal attachments using design of experiments methods. *Journal of Biomechanical Engineering, 128*(3), 399–408.

Yao, J., Snibbe, J., et al. (2006b). Stresses and strains in the medial meniscus of an ACL deficient knee under anterior loading: A finite element analysis with image-based experimental validation. *Journal of Biomechanical Engineering, 128*(1), 135–141.

Chapter 3
Overview of High Tibial Osteotomy and Optimization of the Correction Angle

OA is a degenerative disease of articular cartilage that occurs even in young people and results in pain, swelling, stiffness, a decreased ability to move and, sometimes, the formation of bone spurs (Arokoski et al. 2000; Sharma et al. 2001). Previous studies have revealed that excessive loading on the cartilage can engender damage that could lead to the progression of knee OA (Repo and Finlay 1977; Zhang et al. 1999a, b; Kerin et al. 1998; Quinn et al. 2001; Morel et al. 2006; Chen et al. 2003; Clements et al. 2001; Sharma et al. 2001; Sharma 2001).

Several other researchers have proposed that cartilage damage may be related to stress increase, rather than an absolute magnitude of stress (Akizuki et al. 1986; Appleyard et al. 2003; Setton et al. 1995). Due to abnormal loading conditions at the knee joint, local biomechanical factors can gravely influence the initiation and progression of OA (Sowers 2001; Griffin and Guilak 2005; Englund and Lohmander 2004; Cooper et al. 2000; Arokoski et al. 2000; Sharma et al. 2001).

Subsequently, surgical interventions that slide the loads acting on the damaged tissues have exhibited considerable success in relieving symptoms.

3.1 High Tibial Osteotomy Definition

For a knee afflicted by OA progression in the medial compartment, the HTO is an effective and well-established adjunct to conservative management in an effort to interrupt disease progression. The aim of this surgical procedure is to shift the loading axis from the arthritis medial compartment onto the intact lateral compartment, thereby relieving pain, preventing progression of OA and conventionally delaying the need for total knee replacement through solid preservation of the

---

The original version of this chapter was revised: For detailed information please see Erratum. The erratum to this chapter is available at https://doi.org/10.1007/978-3-319-74158-1_5

© The Author(s), under exclusive licence to Springer International Publishing AG, part of Springer Nature 2018

Z. Trad et al., *FEM Analysis of the Human Knee Joint*, SpringerBriefs in Applied Sciences and Technology, https://doi.org/10.1007/978-3-319-74158-1_3

biological knee structures (Coventry 1984). Different surgical techniques for osteotomies have been described, but the opening wedge HTO and the closing wedge HTO in particular have both been well recognized by surgeons (Coventry 1973; Mina et al. 2008; Michaela et al. 2008; Hui et al. 2011; Insall et al. 1984; Hernigou and Ma 2001; Hernigou et al. 1987). Since correction of the mechanical load axis leads to a regenerative process of the associated compartment, with reduction of subchondral sclerosis (Akamatsu et al. 1997; Takahashi et al. 2002) and spontaneous regeneration of cartilage (van Raaij et al. 2009; Outerbridge 1961; Odenbring et al. 1992; Coventry et al. 1993; FuJISAwA et al. 1979; Koshino et al. 2004), HTO must restore a valgus alignment to the lower extremity in order to provide satisfactory clinical results.

In many recent studies (Akizuki et al. 2008a; Tang and Henderson 2005; Naudie et al. 1999; Billings et al. 2000; Flecher et al. 2006; Gstöttner et al. 2008; Michaela et al. 2008; Hui et al. 2011; Efe et al. 2010; Rinonapoli et al. 1998; Koshino et al. 2004; Sprenger and Doerzbacher 2003), the HTO procedure has been improved due to better patient selection, adequate preoperative planning and improved surgical techniques. However, various factors are believed to be responsible for the failure of HTO over the long term, such as under- or over-correction, advanced age, failure of fixation, patellofemoral arthritis joint instability and lateral thrust. Most such factors can be avoided by ensuring adequate patient selection and preoperative planning, thereby improving surgical techniques. Of those preoperative planning factors, the optimal correction angle is generally considered to be the most critical for the long-term clinical outcome and success of the HTO surgery. Nevertheless, deciding upon the optimal correction angle is a step that has led to much controversy and debate. Poor outcomes, such as OA re-progression and patient dissatisfaction, may occur as a result of a failure to target an optimal desired alignment.

Most previous studies have aimed to establish an optimal range of correction angle empirically based on surgical outcomes, such as survival rate (Akizuki et al. 2008a, b; Korovessis et al. 1999; Koshino et al. 2004; Nakhostine et al. 1993; Rinonapoli et al. 1998; Rudan and Simurda 1990; Sprenger and Doerzbacher 2003; Valenti et al. 1990; Yasuda et al. 1990; Schallberger et al. 2011), knee scoring system (Benzakour et al. 2010; Sprenger and Doerzbacher 2003), cartilage wearing rate (Hernigou et al. 1987), motion analysis (Briem et al. 2007) and radiographical examination (Rudan et al. 1999), while a few studies have correlated the effect of the correction angle to biomechanics, such as in vitro contact pressure on the cartilage (Agneskirchner et al. 2007; Mina et al. 2008) and kinematic analysis (Briem et al. 2007).

3.2 Current FEA Studies on the HTO Procedure

Yang et al. (2010) investigated the effect of the frontal plane tibiofemoral angle on the stress and strain distribution in the knee cartilage during the stance phase of the gait cycle. Within this purpose, 3D FE knee joint models of three healthy subjects with different tibiofemoral angles [0.2° (varus); 7.67° (normal) and 10.34° (valgus)]

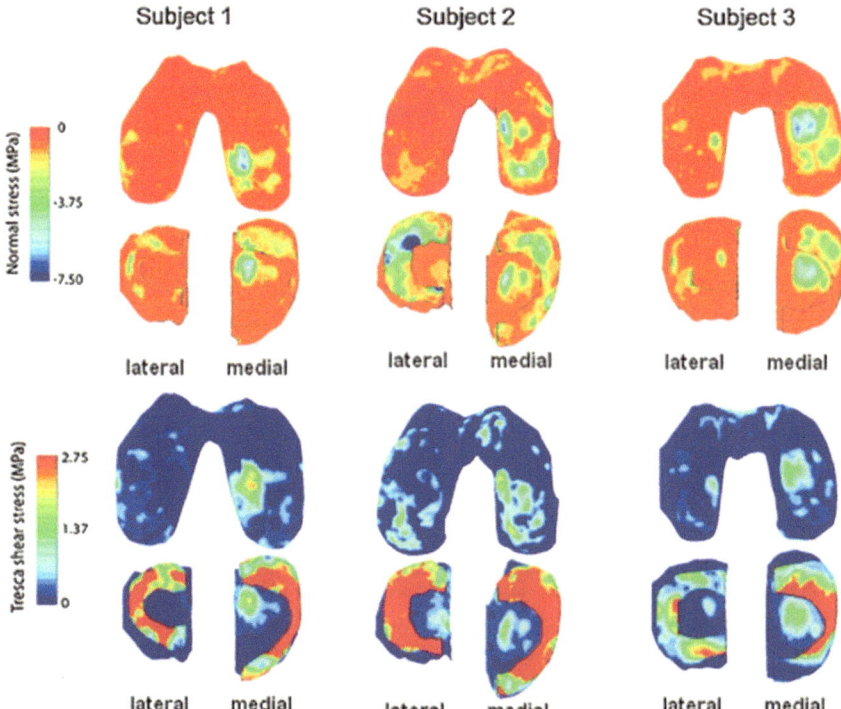

Fig. 3.1 FE results of the compressive stress and tresca shear stress distribution in the knee joint components for subject 1 (varus), subject 2 (normal), and subject 3 (valgus) at approximately 25% of the stance phase of the gait cycle (Yang et al. 2010)

and body weights (640, 725 and 704 N, respectively) were developed based on an MRI of the knee. Loading and boundary conditions were determined from motion analysis and force platform data, in conjunction with the muscle-force reduction method.

As shown in Fig. 3.1, the subject with varus alignment had the largest stresses at the medial compartment of the knee compared to the subjects with normal alignment and valgus alignment, suggesting that this subject might be the most susceptible to developing medial compartment OA. In addition, the magnitude of stress and strain on the lateral cartilage of the subject with valgus alignment were found to be larger compared to subjects with normal alignment and varus alignment, suggesting that this subject might be the most susceptible to developing lateral compartment knee OA.

Despite the small study population and the assumptions included in the models, this study provides useful subject-specific data that explore the role of different biomechanical factors on the stress and strain distribution at the knee cartilage. Thereby, it assesses the effectiveness of OA preventive measures and for long-term follow up studies.

Taking into account its advantages over in vivo cadaveric studies and the potential clinical usefulness in the orthopedic area, the FEA technique was used in a few studies on the HTO procedure. In fact, the study conducted by Zheng (2014) examined the effect of simulated medial open-wedge high tibial osteotomy (OWHTO) with varying correction angles on contact pressure and shear stress distributions within the tibiofemoral cartilages and menisci.

As reported by Jackson et al. (2004), the contact pressure represents the compressive loading on the soft tissues of the knee joint, while the shear stress was believed to be associated with increased catabolic factors and decreased cartilage biosynthetic activity, leading to cartilage damage. A 3D FEM of the human knee joint was developed based on a subject-specific MRI to quantify the stress distribution on the soft tissues of the knee joint under different HTO correction angles of ($0°$, $2.5°$, $5°$, $7.5°$ and $10°$). Bones and articular cartilage were assumed to behave as single-phase linear, elastic and isotropic material, while menisci and ligaments were considered as isotropic and hyperelastic material.

Then, the OWHTO was simulated by bending the proximal portion of the tibia in order to obtain a wedge angle from $0°$ to $10°$ valgus with an increment of $2.5°$, which generated five knee models with different HTO correction degrees of $0°$, $2.5°$, $5°$, $7.5°$ and $10°$, respectively.

The last step was to place the plate, screws and wedge into the cut tibia to simulate the fixation of HTO (Fig. 3.2). Both the screw and fixing plate were created in Solidworks based on the Tomofix HTO plate.

Results of this study showed that both the compressive (Fig. 3.3) and shear stresses (Fig. 3.4) at the cartilage of the medial compartment decrease when the valgus correction angle of the medial opening wedge HTO increases. In addition, the stress distribution at the medial cartilage became more uniform as the loading axis shifted from the medial to the lateral compartment. Therefore, the recommended hip knee ankle (HKA) angle in this study was $6.6°$ of valgus when considering compressive stress and $3.9°$ of valgus when considering shear stress.

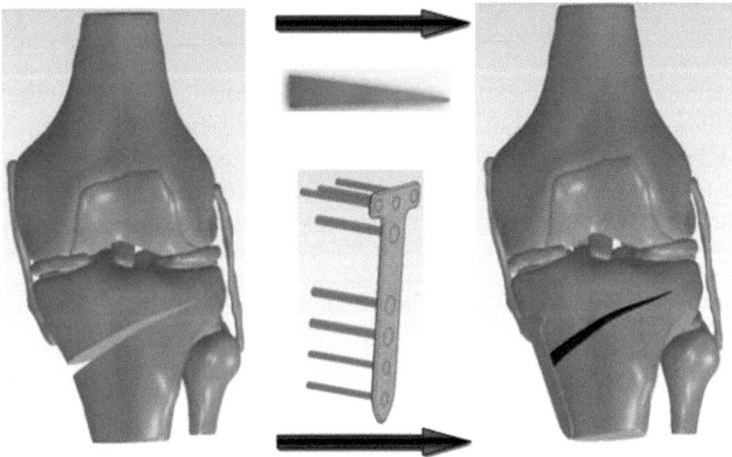

Fig. 3.2 Insertion of the plate, screws and bone-graft substitute (Zheng 2014)

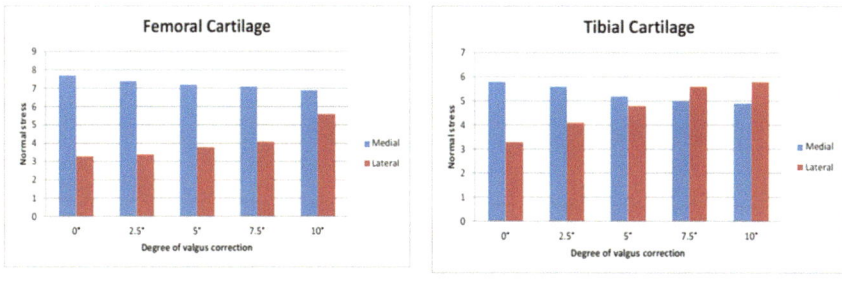

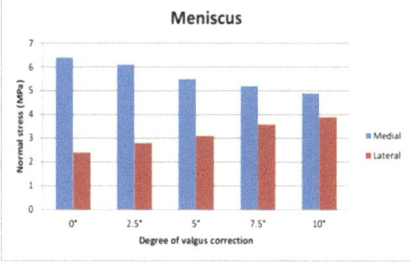

Fig. 3.3 Variation of maximum compressive stresses in the femoral cartilage, tibial cartilage and meniscus of both compartments across the change in valgus correction (Zheng 2014)

These findings are well correlated to the clinical data and recommendations in literature. Moreover, the method adopted in this study could be used as a preoperative assessment tool for predicting the consequential mechanical loading information for surgeons in order to decide upon the optimal patient-specific angle.

Later, Zhu et al. (2015) analyzed the influence of the tibial component coronal alignment on knee biomechanics in mobile-bearing unicompartmental knee arthroplasty (UKA) and determined a ration range of inclination angles. At that time, a 3D FEM of an intact knee was created from the image data of one normal subject. Then, a set of eleven UKA FE models was developed with the coronal inclination angles of the tibial tray ranging from 10° valgus to 10° varus (Fig. 3.5). Tibial bone stresses and strains, contact pressures and load distribution in all UKA models were calculated and analyzed under a compressive load of 1000 N applied to the mid-point of the transepicondylar axis in the femur. The femur was constrained only in flexion-extension, while the tibia and fibula were completely fixed at their distal ends.

As a result, the lateral and medial compartment carried 44.9% and 55.1% of the total load, respectively. The menisci transferred about 69% of the total load. As shown in Fig. 3.6, in UKA models, the von Mises stress and compressive strain at the proximal medial cortical bone increased significantly as the tibial tray was in valgus inclination >4°, which may increase the risk of residual pain. Compressive strains at the tibial keel slot were above the high threshold with varus inclination >4°, which may result in greater risk of component migration. The tibial bone

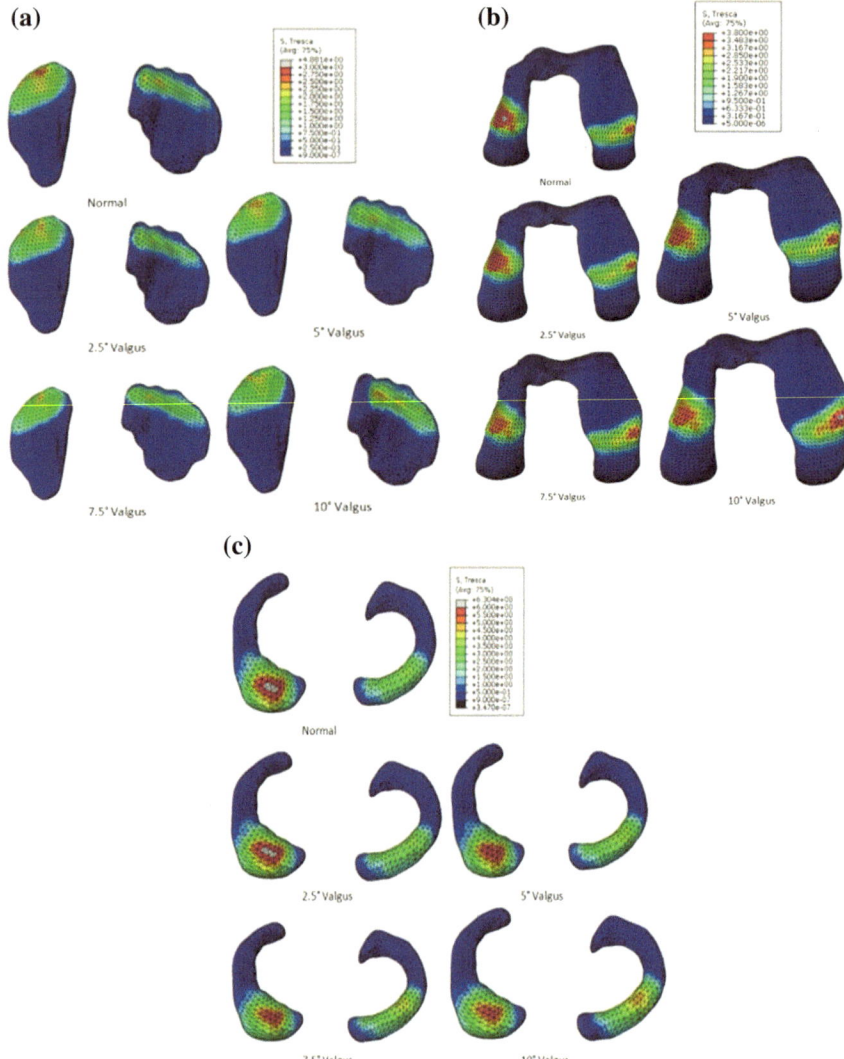

Fig. 3.4 Shear stress distribution in the **a** tibial cartilage, **b** femoral cartilage and **c** menisci of all models (Zheng 2014)

resection corner acted as a strain-raiser regardless of the inclination angles. Compressive strains at the resected surface slightly changed with the varying inclinations and were not supposed to induce bone resorption and component loosening. Finally, contact pressures and load percentage in the lateral compartment increased with the greater varus inclination, which may lead to OA progression. In conclusion, this work suggested that a range of tibial component coronal inclination

Fig. 3.5 Four major FE
models used in analyses
a Intact knee model, **b** UKA
model with neutrally aligned
tibial tray, **c** UKA model with
10° valgus tibial tray, and
d UKA model with 10° varus
tibial tray (Zhu et al. 2015)

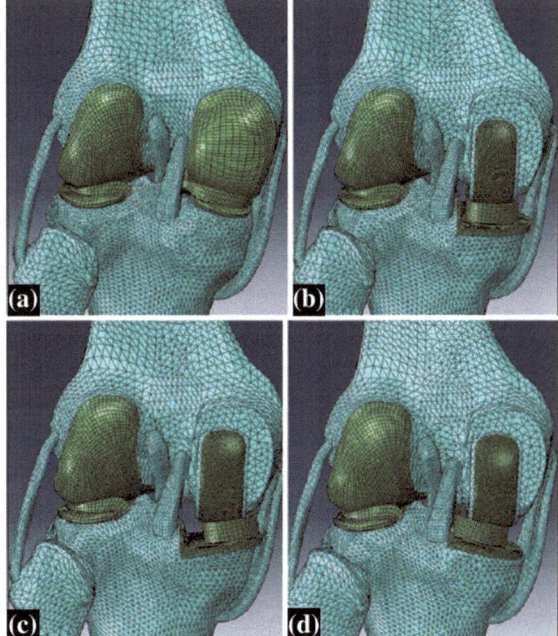

Fig. 3.6 Peak von Mises
stresses and minimum
principal strains at regions of
interest 1 (ROI 1) in all UKA
models (Zhu et al. 2015)

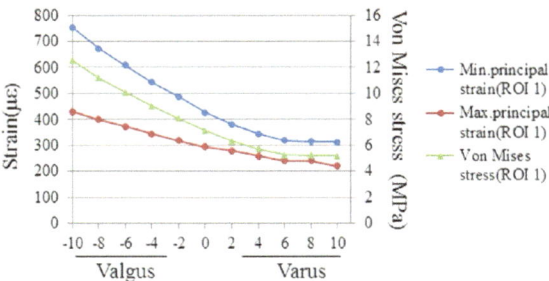

from 4° valgus to 4° varus can be recommended in mobile-bearing UKA for
reducing postoperative complications and enhancing long-term survivorship of
implants as far as possible. Nonetheless, the use of subject-specific data for constructing an FEM of the knee joint and considering bone structures as linear, elastic
and homogeneous materials can affect the extrapolation of the results. In addition,
an accurate study of stress and strain distribution may be performed under a
dynamic simulation of the knee joint at varying knee flexion angles.

On the other hand, Blecha et al. (2005) evaluated the effect of fixing the supporting
plate (SP) in an anteromedial position rather than in a medial position on the stress
distribution and micromotion magnitude. A 3D FEM was developed to investigate the
OWHTO numerically. Based on the study by Hernigou and Ma (2001), the wedge

Fig. 3.7 Puddu plate positioning (**1**) and Tomofix plate positioning (**2**) on 3D models of simulated OWHTO (Izaham et al. 2012)

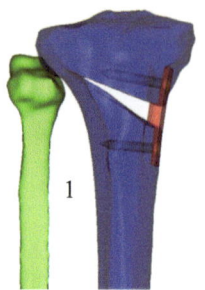 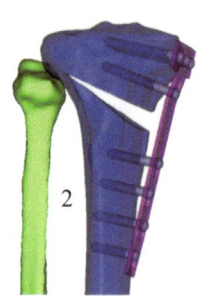

was assumed to be made of acrylic surgical cement and was placed in a posteromedial position that corresponds to the current clinical practice. Findings of this work revealed that, under maximal gait load, the maximal stresses observed in the four structural elements (bone, plate, wedge, screws) of an osteotomy with a plate in the medial position were substantially three times higher than those observed in an osteotomy with an anteromedial plate configuration. For this reason, positioning the SP medially was found to support smaller loading than an OWHTO with an anteromedial plate position, thus achieving smaller structural stability and safety. Furthermore, positioning the plate medially presented a greater magnitude of micromotions at the bone/wedge interface than positioning it in the anteromedial position. Finally, an OWHTO with an anteromedial plate should be loaded under 18.8% of the normal gait load until the union of osteotomy interfaces is achieved.

Another study conducted by Izaham et al. (2012) analyzed the rigidity and stability of the Puddu and Tomofix plates, which are the two most common plates, in securing the osteotomized site in OWHTO procedures. In fact, a 3D FEM of the tibia was created in order to simulate an osteotomy with the fixation of the two plates (Fig. 3.7). The model was fixed distally in all degrees of freedom. Simulated data generated from the micromotions, displacement and implant stress were captured.

As shown in Fig. 3.8, the Tomofix fixation produced high concentrated stress at the lateral hinge of the osteotomy. The micromotion analysis (Fig. 3.9) demonstrates that the Puddu plate underwent higher amounts of motion than the Tomofix plates. In addition, higher micromotion was observed on the surface of the Puddu plate, which was in contact with the bone, especially at the distal part of the plate. To conclude, while this study has demonstrated that the use of Tomofix plates provides better rigidity, and therefore stability, than the Puddu plate system, the stress experienced in the Tomofix plates may nevertheless be of concern, especially in view of the possibility that this implant may fail. In addition, this study demonstrated that a precise HTO simulation involving bending of the tibia to vary lower limb alignment would definitely provide more realistic and clinically relevant results.

Fig. 3.8 Equivalent von Mises stress distribution along the model: Bottom view of plates (**a**, **e**); plates side view (**b**, **f**); anterior view of fixation models (**c**, **d**), (**g**, **h**) (Izaham et al. 2012)

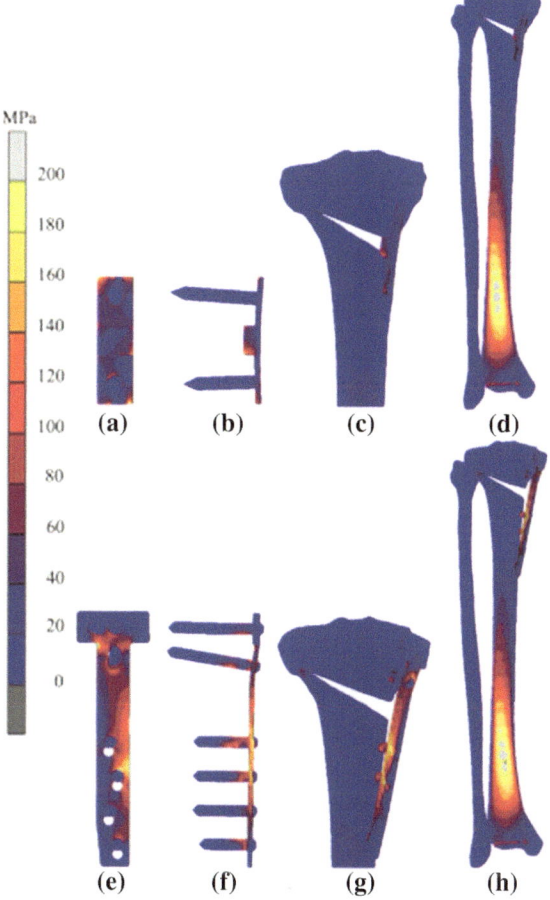

3.3   Current Clinical Studies on Optimizing the Correction Angle

Achieving the appropriate alignment is a critical factor of success for an HTO in treating OA of either the medial or lateral compartment of the knee joint. However, it is surprising, after several decades of osteotomy, that there is little agreement between studies on the ideal alignment. In some studies (El-Azab et al. 2011; Takeuchi et al. 2009; Schallberger et al. 2011), the purpose is a lateral shift of the weight-bearing line to a maximum of 62.5% of the medial to lateral width of the tibia plateau, which is known as the Fujisawa point (FuJISAwA et al. 1979), equivalent to 3.5° of valgus (Rudan et al. 1999). The precise position of the correction was dependent upon the magnitude of mal-alignment and the status of the articular cartilage in the lateral tibiofemoral compartment as evaluated through

Fig. 3.9 Micromotion result for the Tomofix plate (**a, b**) and Puddu's plate (**c, d**) (Izaham et al. 2012)

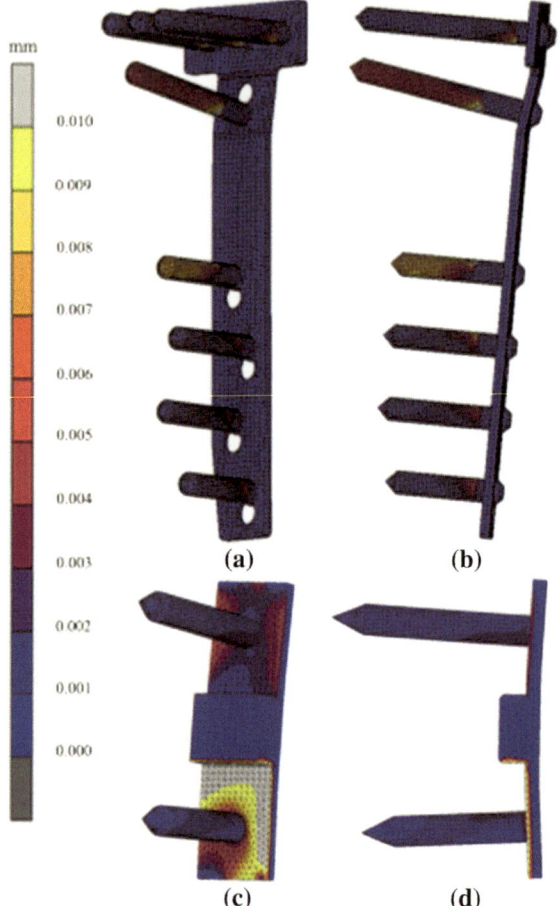

intraoperative arthroscopy. For those patients with relatively healthy cartilage in the lateral compartment, an HKA angle of 3.5° valgus was sought; otherwise, the aim was towards 0°. However, after a large number of clinical studies targeting the ideal correction angle, the recommended angles now vary from 1° valgus up to 10° valgus.

In order to better understand the up-to-date investigation of optimal alignment following HTO, Table 3.1 summarises all of the published clinical studies available in literature with a focus on an ideal post-operative alignment. Despite the fact that some of them used the anatomical axis rather than the mechanical axis to measure the alignment, it is easy to convert between mechanical alignment (also known as the HKA angle) and anatomical alignment by considering an offset of 5° between them.

We note that among these studies, the advisable optimal alignment following HTO varies significantly by author; some of them even have contradictory results.

Table 3.1 Summary of up-to-date clinical studies in literature on the optimal correction angle

Author(s)/Reference	Year	Mechanical alignment (°)	Anatomical alignment (°)
Coventry (1984)	1984	10	
Insall et al. (1984)	1984		5–14
Hernigou et al. (1987)	1987	3–6	
Valenti et al. (1990)	1990	3–8	
Hsu et al. (1990)	1990	4.2	
Yasuda et al. (1990)	1990		12–16
Nakhostine et al. (1993)	1993		5–7
Rinonapoli et al. (1998)	1998		10–12
Korovessis et al. (1999)	1999		6–10
Rudan et al. (1999)	1999		10.8
Aglietti et al. (2003)	2003		8–15
Sprenger and Doerzbacher (2003)	2003		8–16
Koshino et al. (2004)	2004		9
Takeuchi et al. (2009)	2009		10.4
Birmingham et al. (2009)	2009	1	
Benzakour et al. (2010)	2010	5	

In the study by Hernigou et al. (1987), the results were elucidated in a manner that is somewhat self-confirming, and while the 3°–6° range may be a good target for post-operative alignment at follow-up, greater over-correction in surgery is ideal for underwriting that this desired range is actually reached and maintained by the majority of knees in the long term. Furthermore, this narrow range fulfills most of the clinical studies in the literature. The study presented by Benzakour et al. (2010) is the most recent clinical study investigating the optimal correction in HTO. In their work, an average 15-year follow-up on 224 knees with HTO treatments has been conducted, in which the excellent-to-good results were 42% in the group with an average mechanical valgus of 5°, which can be quite effective at the follow-up for at least ten years. Unfortunately, however, determination of the ideal HTO correction angle remains controversial.

3.4 Current Biomechanical Studies on Optimizing the Correction Angle

Other than merely studying the effect of the amount of frontal plane knee alignment correction following HTO on clinical outcome, Briem et al. (2007) investigated the effect of the degree of frontal plane knee alignment following OWHTO surgery on joint moments, muscle co-contraction and self-reported functional outcome. However, the results lacked survey and precision, since they used the mean correction angles to define the under-correction and over-correction groups and found that over-correction led to a high adduction moment and muscle co-contraction, which are associated with poor HTO survival rate. Despite the reduction of the optimal correction angle to smaller than 5° valgus, the deficiency of bringing back the kinematics and kinetic findings to the knee joint's internal loading limited this study to further exploring the effect of changing alignment on a knee joint's internal loading, as the actual loading on the cartilage and menisci is more relative to OA progression.

On the other hand, Mina et al. (2008) tried to directly relate the correction angles to the contact loading pattern of the knee joint by using an electronic pressure sensor inserted into the cartilage of a cadaveric knee specimen under a specific mechanical testing system simulating functional activity of the knee joint (Fig. 3.10). Eight human cadaver knees had been used to perform HTOs. Each leg was tested through 10 different adduction angles using an osteotomy spreader in replication of a real HTO treatment. Aiming to achieve equal stress distribution between the medial and lateral compartments of the knee joint, which is considered to most closely approximate idealised physiologic loading, they (Mina et al. 2008) recommended an alignment of 0°–4° of valgus (Fig. 3.11), regardless of condylar width, baseline tibiofemoral alignment, body weight and the size of the chrondral defect. Although this recommendation has reduced the optimal HTO alignment somewhat, it has enfolded one of the major challenges in HTO research: the fact that the in vivo measurement of stress and strain at the knee cartilage following HTO treatment is arduous during a dynamic loading cycle.

Further work has been conducted by Agneskirchner et al. (2007) in order to quantify the effect of different loading axes, as well as a valgus OWHTO, on tibiofemoral cartilage pressure. Six human knee specimens were tested under a compressive load of 1000 N applied with a hydraulic apparatus. Then, the increased pressure in the medial joint compartment at 8° valgus was observed using pressure-sensitive films. Several results were drawn from this study. First, the position of the loading axis in the frontal plane has a strong effect on the tibiofemoral cartilage pressure distribution of the knee. Then, the medial compartment is

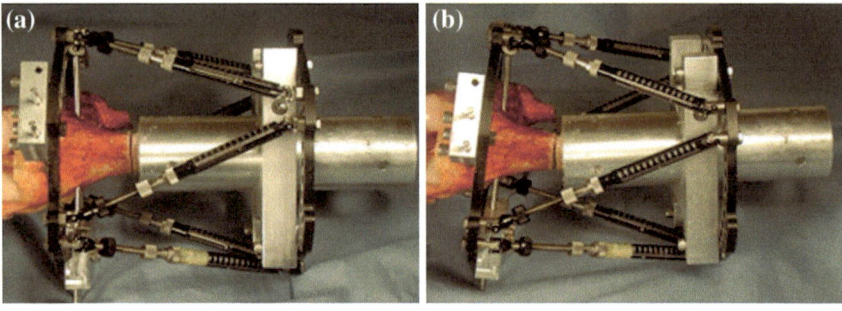

Fig. 3.10 High Tibial Osteotomy within the Taylor Spatial Frame before (**a**) and after (**b**) medial wedge opening to 10° of valgus (Mina et al. 2008)

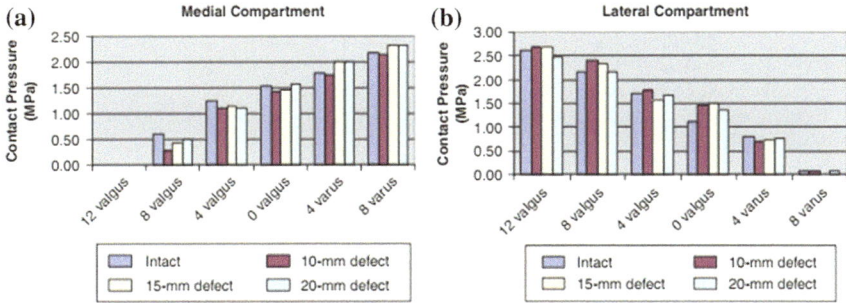

Fig. 3.11 Average contact pressure for each defect size at various tibiofemoral alignments within the medial (**a**) and lateral (**b**) compartments (Mina et al. 2008)

predominantly loaded in a varus knee, while a neutral mechanical axis loads the lateral slightly more than the medial compartment. Finally, in valgus alignment, the leading load runs through the lateral compartment. The clinical pertinence carried by this study was that an OWHTO maintains high medial compartment pressure despite the fact that the loading axis has been shifted into valgus. Only after complete release of the distal fibers of the MCL does the opening wedge HTO produce a decompression of the medial joint compartment.

The reason for the discrepancy between this study and the study by Mina et al. (2008) may be due to differences in the location of the osteotomy. In their study, the osteotomy was superior to the distal attachment region of the MCL, which led to a superiorly directed force on the medial tibial plateau, because of ligament tension caused by the MCL.

References

Aglietti, P., et al. (2003). High tibial valgus osteotomy for medial gonarthrosis: a 10-to 21-year study. *The Journal of Knee Surgery, 16*(1), 21–26.

Agneskirchner, J.D. et al., 2007. The effects of valgus medial opening wedge high tibial osteotomy on articular cartilage pressure of the knee: A biomechanical study. *Arthroscopy: The Journal of Arthroscopic & Related Surgery, 23*(8), 852–861.

Akamatsu, Y., et al. (1997). Changes in osteosclerosis of the osteoarthritic knee after high tibial osteotomy. *Clinical Orthopaedics and Related Research, 334,* 207–214.

Akizuki, S., et al. (1986). Tensile properties of human knee joint cartilage: I. Influence of ionic conditions, weight bearing, and fibrillation on the tensile modulus. *Journal of Orthopaedic Research, 4*(4), 379–392.

Akizuki, S., et al. (2008a). The long-term outcome of high tibial osteotomy. *Bone & Joint Journal, 90*(5), 592–596.

Akizuki, S., et al. (2008b). The long-term outcome of high tibial osteotomy A TEN-TO 20-YEAR FOLLOW-UP. *Journal of Bone and Joint Surgery. British Volume, 90*(5), 592–596.

Appleyard, R. C., et al. (2003). Topographical analysis of the structural, biochemical and dynamic biomechanical properties of cartilage in an ovine model of osteoarthritis. *Osteoarthritis and Cartilage, 11*(1), 65–77.

Arokoski, J. P. A., et al. (2000). Normal and pathological adaptations of articular cartilage to joint loading. *Scandinavian Journal of Medicine and Science in Sports, 10*(4), 186–198.

Benzakour, T. et al. (2010). h. *International orthopaedics, 34*(2), 209–215.

Billings, A., et al. (2000). High tibial osteotomy with a calibrated osteotomy guide, rigid internal fixation, and early motion. Long-term follow-up. *Journal of Bone and Joint Surgery. American Volume, 82*(1), 70–79.

Birmingham, T. B., et al. (2009). Medial opening wedge high tibial osteotomy: A prospective cohort study of gait, radiographic, and patient-reported outcomes. *Arthritis Care & Research, 61*(5), 648–657.

Blecha, L. D., et al. (2005). How plate positioning impacts the biomechanics of the open wedge tibial osteotomy; a finite element analysis. *Computer Methods in Biomechanics and Biomedical Engineering, 8*(5), 307–313.

Briem, K., et al. (2007). Effects of the amount of valgus correction for medial compartment knee osteoarthritis on clinical outcome, knee kinetics and muscle co-contraction after opening wedge high tibial osteotomy. *Journal of Orthopaedic Research, 25*(3), 311–318.

Chen, C., et al. (2003). Time, stress, and location dependent chondrocyte death and collagen damage in cyclically loaded articular cartilage. *Journal of Orthopaedic Research, 21*(5), 888–898.

Clements, K. M., et al. (2001). How severe must repetitive loading be to kill chondrocytes in articular cartilage? *Osteoarthritis and Cartilage, 9*(5), 499–507.

Cooper, C., et al. (2000). Risk factors for the incidence and progression of radiographic knee osteoarthritis. *Arthritis and Rheumatism, 43*(5), 995–1000.

Coventry, M. B. (1973). Osteotomy about the knee for degenerative and rheumatoid arthritis. *Journal of Bone and Joint Surgery. American Volume, 55*(1), 23–48.

Coventry, M. B. (1984). Upper tibial osteotomy. *Clinical Orthopaedics and Related Research, 182*, 46–52.

Coventry, M. B., Ilstrup, D. M., & Wallrichs, S. L. (1993). Proximal tibial osteotomy. A critical long-term study of eighty-seven cases. *Journal of Bone and Joint Surgery, 75*(2), 196–201.

Efe, T., et al. (2010). TKA following high tibial osteotomy versus primary TKA-a matched pair analysis. *BMC musculoskeletal disorders, 11*(1), 207.

El-Azab, H. M., et al. (2011). Limb alignment after open-wedge high tibial osteotomy and its effect on the clinical outcome. *Orthopedics, 34*(10), e622–e628.

Englund, M., & Lohmander, L. S. (2004). Risk factors for symptomatic knee osteoarthritis fifteen to twenty-two years after meniscectomy. *Arthritis and Rheumatism, 50*(9), 2811–2819.

Flecher, X., et al. (2006). A 12-28-year followup study of closing wedge high tibial osteotomy. *Clinical Orthopaedics and Related Research, 452*, 91–96.

FuJISAwA, Y., Masuhara, K., & Shiomi, S. (1979). The effect of high tibial osteotomy on osteoarthritis of the knee. An arthroscopic study of 54 knee joints. *The Orthopedic clinics of North America, 10*(3), 585–608.

Griffin, T. M., & Guilak, F. (2005). The role of mechanical loading in the onset and progression of osteoarthritis. *Exercise and Sport Sciences Reviews, 33*(4), 195–200.

Gstöttner, M., et al. (2008). Long-term outcome after high tibial osteotomy. *Archives of Orthopaedic and Trauma Surgery, 128*(3), 345.

Hernigou, P.H. et al. (1987). Proximal tibial osteotomy for osteoarthritis with varus deformity. A ten to thirteen-year follow-up study. *The Journal of bone and joint surgery. American volume, 69*(3), 332–354.

Hernigou, P., & Ma, W. (2001). Open wedge tibial osteotomy with acrylic bone cement as bone substitute. *The Knee, 8*(2), 103–110.

Hsu, R. W. W., et al. (1990). Normal axial alignment of the lower extremity and load-bearing distribution at the knee. *Clinical Orthopaedics and Related Research, 255*, 215–227.

Hui, C., et al. (2011). Long-term survival of high tibial osteotomy for medial compartment osteoarthritis of the knee. *The American journal of sports medicine, 39*(1), 64–70.

Insall, J. N., Joseph, D. M., & Msika, C. (1984). High tibial osteotomy for varus gonarthrosis. A long-term follow-up study. *Journal of Bone and Joint Surgery. American Volume, 66*(7), 1040–1048.

Izaham, R. M. A. R., et al. (2012). Finite element analysis of Puddu and Tomofix plate fixation for open wedge high tibial osteotomy. *Injury, 43*(6), 898–902.

Jackson, B. D., et al. (2004). The effect of the knee adduction moment on tibial cartilage volume and bone size in healthy women. *Rheumatology, 43*(3), 311–314.

Kerin, A. J., Wisnom, M. R., & Adams, M. A. (1998). The compressive strength of articular cartilage. *Proceedings of the Institution of Mechanical Engineers. Part H, Journal of Engineering in Medicine, 212*(4), 273–280.

Korovessis, P., et al. (1999). Medium-and long-term results of high tibial osteotomy for varus gonarthrosis in an agricultural population. *Orthopedics, 22*(8), 729–735.

Koshino, T., et al. (2004). Fifteen to twenty-eight years' follow-up results of high tibial valgus osteotomy for osteoarthritic knee. *The Knee, 11*(6), 439–444.

Michaela, G., et al. (2008). Long-term outcome after high tibial osteotomy. *Archives of Orthopaedic and Trauma Surgery, 128*(1), 111–115.

Mina, C., et al. (2008). High tibial osteotomy for unloading osteochondral defects in the medial compartment of the knee. *The American journal of sports medicine, 36*(5), 949–955.

Morel, V., Berutto, C., & Quinn, T. M. (2006). Effects of damage in the articular surface on the cartilage response to injurious compression in vitro. *Journal of Biomechanics, 39*(5), 924–930.

Nakhostine, M., et al. (1993). A special high tibial osteotomy technique for treatment of unicompartmental osteoarthritis of the knee. *Orthopedics, 16*(11), 1255–1258.

Naudie, D., et al. (1999). Survivorship of the High Tibial Valgus Osteotomy A 10-to 22-Year Followup Study. *Clinical Orthopaedics and Related Research, 367,* 18–27.

Odenbring, S., et al. (1992). Cartilage regeneration after proximal tibial osteotomy for medial gonarthrosis: an arthroscopic, roentgenographic, and histologic study. *Clinical Orthopaedics and Related Research, 277,* 210–216.

Outerbridge, R.E. (1961). The etiology of chondromalacia patellae. *Journal of Bone and Joint Surgery British, 43.*

Quinn, T. M., et al. (2001). Matrix and cell injury due to sub-impact loading of adult bovine articular cartilage explants: effects of strain rate and peak stress. *Journal of Orthopaedic Research, 19*(2), 242–249.

Repo, R. U., & Finlay, J. B. (1977). Survival of articular cartilage after controlled impact. *Journal of Bone and Joint Surgery. American Volume, 59*(8), 1068–1076.

Rinonapoli, E., et al. (1998). Tibial Osteotomy for Varus Gonarthrosis: A 10-to 21-Year Followup Study. *Clinical Orthopaedics and Related Research, 353,* 185–193.

Rudan, J. F., & Simurda, M. A. (1990). High Tibial Osteotomy: A Prospective Clinical and Roentgenographic Review. *Clinical Orthopaedics and Related Research, 255,* 251–256.

Rudan, J., Harrison, M., & Simurda, M. A. (1999). Optimizing femorotibial alignment in high tibial osteotomy. *Canadian Journal of Surgery, 42*(5), 366.

Schallberger, A., et al. (2011). High tibial valgus osteotomy in unicompartmental medial osteoarthritis of the knee: A retrospective follow-up study over 13–21 years. *Knee Surgery, Sports Traumatology, Arthroscopy, 19*(1), 122–127.

Setton, L. A., Mow, V. C., & Howell, Ds. (1995). Mechanical behavior of articular cartilage in shear is altered by transection of the anterior cruciate ligament. *Journal of Orthopaedic Research, 13*(4), 473–482.

Sharma, L. (2001). Local factors in osteoarthritis. *Current Opinion in Rheumatology, 13*(5), 441–446.

Sharma, L., et al. (2001). The role of knee alignment in disease progression and functional decline in knee osteoarthritis. *JAMA, 286*(2), 188–195.

Sowers, M. (2001). Epidemiology of risk factors for osteoarthritis: systemic factors. *Current Opinion in Rheumatology, 13*(5), 447–451.

Sprenger, T. R., & Doerzbacher, J. F. (2003). Tibial osteotomy for the treatment of varus gonarthrosis. *Journal of Bone and Joint Surgery. American Volume, 85*(3), 469–474.

Takahashi, S., Koshino, T., & Saito, T. (2002). Decrease of osteosclerosis in subchondral bone of medial compartmental osteoarthritic knee seven to nineteen years after high tibial valgus osteotomy. *Bulletin-Hospital for Joint Diseases, 61*(1–2), 58–62.

Takeuchi, R. et al. (2009). Medial opening wedge high tibial osteotomy with early full weight bearing. *Arthroscopy: The Journal of Arthroscopic & Related Surgery, 25*(1), 46–53.

Tang, W. C., & Henderson, I. J. P. (2005). High tibial osteotomy: long term survival analysis and patients' perspective. *The Knee, 12*(6), 410–413.

Valenti, J. R., et al. (1990). Long term evaluation of high tibial valgus osteotomy. *International Orthopaedics, 14*(4), 347–349.

van Raaij, T.M. et al. (2009). Survival of closing-wedge high tibial osteotomy: Good outcome in men with low-grade osteoarthritis after 10–16 years. *Acta orthopaedica, 79*(2), 230–234.

Yang, N. H., et al. (2010). Effect of frontal plane tibiofemoral angle on the stress and strain at the knee cartilage during the stance phase of gait. *Journal of Orthopaedic Research, 28*(12), 1539–1547.

Yasuda, K., et al. (1990). Long-term evaluation of high tibial osteotomy for medial osteoarthritis of the knee. *Bulletin of the Hospital for Joint Diseases Orthopaedic Institute, 51*(2), 236–248.

Zhang, H. et al. (1999a). Magnetic Resonance Image Based 3D Poroelastic Finite Element Model of Tibio-Menisco-Femoral Contact. In *23rd Proceedings of the American Society of Biomechanics*, 198–199.

Zhang, H., et al. (1999b). Damage to rabbit femoral articular cartilage following direct impacts of uniform stresses: An in vitro study. *Clinical Biomechanics, 14*(8), 543–548.

Zheng, K. (2014). The effect of High Tibial Osteotomy correction angle on Cartilage and Meniscus loading using finite element analysis.

Zhu, G.-D., et al. (2015). Finite element analysis of mobile-bearing unicompartmental knee arthroplasty: The influence of tibial component coronal alignment. *Chinese Medical Journal, 128*(21), 2873.

Chapter 4
Conclusions and Future Work

In this manuscript, we have provided an overview of the overall biomechanical and clinical studies on healthy and pathological knee joints. The various methods used by different researchers for modeling this joint were evoked, as well as the material properties assigned for each component of the knee joint. Ultimately, the different experimental and FEA studies that investigate the mechanical behavior of the knee joint, as well as the cause, treatment and prevention of knee OA, were well analyzed in order to discuss its major effect on the clinical area. This review discussed various FEA studies available in literature that enable a biomechanical analysis of knee behavior after a specific treatment (ligament rupture, meniscectomy, HTO, etc.), and provided clinically relevant information regarding the force, stress and displacement changes that occur in a pathological knee. Ultimately, the work presented here is a new concept focused on the FEA method that takes a step towards a novel diagnostic tool for the assessment of possible failure sites in the human knee, assessing the outcome of treatments or surgical procedures, such as the HTO, and that could thus be used in clinical decision-making.

The main conclusions of these studies can be summarized as follows:

- 3D FEA knee models have been used to analyze the contact behavior at the knee joint, with respect to different parameters, boundary conditions, and constraint analysis, ligament injury or deficiency and meniscectomy.
- Accurate geometries of the bony structure and soft tissues of the human knee joint are created with the help of computed tomography and magnetic resonance imaging scanners. Some existing FEA studies created 3D knee geometries from cadavers and some from MRI of live human subjects.
- The different forces and torques on the knee joint can be obtained from motion photography and force transducer.
- The majority of 3D computational knee gait models have employed cadaver data to define the geometry and data from a live person to define the kinematics and kinetics of the knee joint.

© The Author(s), under exclusive licence to Springer International Publishing AG, part of Springer Nature 2018
Z. Trad et al., *FEM Analysis of the Human Knee Joint*, SpringerBriefs in Applied Sciences and Technology, https://doi.org/10.1007/978-3-319-74158-1_4

- The FE simulation technique allows for precise computation of both special and temporal variations of stresses, strains and contact areas/forces in different situations that can be easily reproduced. Thus, FEA and modeling techniques are a powerful tool in providing biomechanical information that can be extremely helpful in a clinical context. These procedures play a vital role in human knee analysis and the design of artificial knee implants.
- The FEA method can provide valuable information that may be utilized by medical professionals, including physicians, rehabilitation specialists, nutritionists and bioengineers, to develop OA preventive measures, rehabilitation and surgical procedures.
- It has been widely reported that the force distribution at the knee joint is correlated to the tibiofemoral angle and varus knee moment.
- HTO has seen widespread use in the last half century, since optimising the correction angle of medial OWHTO pre-operatively based on patient-specific knee anatomy and subsequent biomechanics may increase survival rate of the procedure and restoration of post-operative patient function.
- Regardless of the critical aspect to choosing the optimal angle of valgus correction to achieve the long-term survival result, several studies on clinical outcome have been related to pursuit of such an empirical optimal angle, with as many conclusions drawn as there are authors. Most of the above studies were largely clinically-based and lacked biomechanical insights into the causes of such controversial recommendations.
- Although the range of recommended correction degrees has been minimized through cadaveric mechanical testing, which directly explored the effect of changing alignment and loading in cartilage, the necessity of inserting electronic pressure sensors, particularly under the menisci, limits this approach to in vivo studies.
- The FEA method could be used to identify those people most susceptible to OA, develop preventive measures and be employed for long-term follow-up. It addresses the need to determine the knee cartilage stress and strain based on subject specific loading conditions and geometry.
- Considering the high costs associated with OA, it is a relatively inexpensive method, and with the cost of MRI becoming cheaper and the use of computational biomechanics becoming better, with faster, stronger computational abilities, it has the potential to become a tool that can be used in the clinical setting to give assistance to subject-specific OA prevention methods.

Facing the key challenge of quantifying the effects of varying correction angles on patient-specific knee load distribution non-invasively prior to surgery, it is worthwhile involving a new approach targeting optimisation of the HTO alignment based on knee biomechanics preoperatively. This process should be capable of providing insightful understandings comparable to those derived from the experiments of mechanical testing or follow-up assessment of clinical studies. Unfortunately, there have been no studies available on assessment of the effect of variation in knee alignment, in terms of the correction angle, using FEA. Consequently, with an improved FEA knee model that includes the patellofemoral joint and subject-specific muscle architecture, the stresses at the knee cartilage could be found to determine the differences between healthy and pathological individuals.

Moreover, the modeling process has several issues that future studies will need to resolve. These are, for instance: a reliable automatic segmentation of tissues from MR images, soft tissue properties (ligament cartilage and menisci), patient-specific physiological loading conditions, as well a selection of different functional activities.

Although gait is the default physical activity to be simulated, the research will eventually spread out to other activities, such as stair climbing, lunging and crouching.

Furthermore, as regards the optimization of the correction angle, repeating the investigation into a number of real patients operated on with clinical HTO procedures will allow for a better comparison of results. Acquisition of greater biomechanical and clinical insights will be expected through determination as to whether similar stress distributions are obtained across different patients and whether the same patterns of stress distribution would exist in different patients.

In summary, the overall kinematics and biomechanics of the knee have been studied extensively and significant advances have been made in this area. However, due to the complexity of the knee joint structure and the complexity of defining the accurate mechanical properties for the radial, transition and tangential cartilage zones, and the lack of consensus on the optimal correction degrees for HTO, numerous obstacles remain. These make this clinically important problem challenging and still very open for further scientific investigation.

Erratum to: Overview of High Tibial Osteotomy and Optimization of the Correction Angle

Zahra Trad, Abdelwahed Barkaoui, Moez Chafra and João Manuel R. S. Tavares

Erratum to:
Chapter 3 in: Z. Trad et al., *FEM Analysis of the Human Knee Joint*, SpringerBriefs in Applied Sciences and Technology, https://doi.org/10.1007/978-3-319-74158-1_3

In the original version of the book, the belated corrections from author have to be incorporated in Chap. 3:

In Table 3.1, the citation "Valentini et al." has to be corrected as "Valenti et al.", and the values 4–9, 8–14 and 2–4 have to be changed as 4.2, 8–15 and 5, respectively.

The sentence "In their work, … in the group with an angle less than 4°." has to be replaced with "In their work, …in the group with an average mechanical valgus of 5°, which can be quite effective at the follow-up for at least ten years."

The erratum chapter and the book have been updated with the changes.

The updated original online version for this chapter can be found at https://doi.org/10.1007/978-3-319-74158-1_3

© The Author(s), under exclusive licence to Springer International Publishing AG, part of Springer Nature 2018
Z. Trad et al., *FEM Analysis of the Human Knee Joint*, SpringerBriefs in Applied Sciences and Technology, https://doi.org/10.1007/978-3-319-74158-1_5